INTRODUCTION TO MATHEMATICAL PHILOSOPHY

By BERTRAND RUSSELL

A Digireads.com Book
Digireads.com Publishing

Introduction to Mathematical Philosophy
By Bertrand Russell
ISBN: 1-4209-3840-1

Please visit *www.digireads.com*

CONTENTS

PREFACE

This book is intended essentially as an "Introduction," and does not aim at giving an exhaustive discussion of the problems with which it deals. It seemed desirable to set forth certain results, hitherto only available to those who have mastered logical symbolism, in a form offering the minimum of difficulty to the beginner. The utmost endeavour has been made to avoid dogmatism on such questions as are still open to serious doubt, and this endeavour has to some extent dominated the choice of topics considered. The beginnings of mathematical logic are less definitely known than its later portions, but are of at least equal philosophical interest. Much of what is set forth in the following chapters is not properly to be called "philosophy," though the matters concerned were included in philosophy so long as no satisfactory science of them existed. The nature of infinity and continuity, for example, belonged in former days to philosophy, but belongs now to mathematics. Mathematical philosophy, in the strict sense, cannot, perhaps, be held to include such definite scientific results as have been obtained in this region; the philosophy of mathematics will naturally be expected to deal with questions on the frontier of knowledge, as to which comparative certainty is not yet attained. But speculation on such questions is hardly likely to be fruitful unless the more scientific parts of the principles of mathematics are known. A book dealing with those parts may, therefore, claim to be an introduction to mathematical philosophy, though it can hardly claim, except where it steps outside its province, to be actually dealing with a part of philosophy. It does deal, however, with a body of knowledge which, to those who accept it, appears to invalidate much traditional philosophy, and even a good deal of what is current in the present day. In this way, as well as by its bearing on still unsolved problems, mathematical logic is relevant to philosophy. For this reason, as well as on account of the intrinsic importance of the subject, some purpose may be served by a succinct account of the main results of mathematical logic in a form requiring neither a knowledge of mathematics nor an aptitude for mathematical symbolism. Here, however, as elsewhere, the method is more important than the results, from the point of view of further research; and the method cannot well be explained within the framework of such a book as the following. It is to be hoped that some readers may be sufficiently interested to advance to a study of the method by which mathematical logic can be made helpful in investigating the traditional problems of philosophy. But that is a topic with which the following pages have not attempted to deal.

BERTRAND RUSSELL.

EDITOR'S NOTE

[The note below was written by J. H. Muirhead, LL.D., editor of the Library of Philosophy series in which Introduction to Mathematical Philosophy was originally published.]

Those who, relying on the distinction between Mathematical Philosophy and the Philosophy of Mathematics, think that this book is out of place in the present Library, may be referred to what the author himself says on this head in the Preface. It is not necessary to agree with what he there suggests as to the readjustment of the field of philosophy by the transference from it to mathematics of such problems as those of class, continuity, infinity, in order to perceive the bearing of the definitions and discussions that follow on the work of "traditional philosophy." If philosophers cannot consent to relegate the criticism of these categories to any of the special sciences, it is essential, at any rate, that they should know the precise meaning that the science of mathematics, in which these concepts play so large a part, assigns to them. If, on the other hand, there be mathematicians to whom these definitions and discussions seem to be an elaboration and complication of the simple, it may be well to remind them from the side of philosophy that here, as elsewhere, apparent simplicity may conceal a complexity which it is the business of somebody, whether philosopher or mathematician, or, like the author of this volume, both in one, to unravel.

INTRODUCTION TO MATHEMATICAL PHILOSOPHY

CHAPTER I

THE SERIES OF NATURAL NUMBERS

Mathematics is a study which, when we start from its most familiar portions, may be pursued in either of two opposite directions. The more familiar direction is constructive, towards gradually increasing complexity: from integers to fractions, real numbers, complex numbers; from addition and multiplication to differentiation and integration, and on to higher mathematics. The other direction, which is less familiar, proceeds, by analysing, to greater and greater abstractness and logical simplicity; instead of asking what can be defined and deduced from what is assumed to begin with, we ask instead what more general ideas and principles can be found, in terms of which what was our starting-point can be defined or deduced. It is the fact of pursuing this opposite direction that characterizes mathematical philosophy as opposed to ordinary mathematics. But it should be understood that the distinction is one, not in the subject matter, but in the state of mind of the investigator. Early Greek geometers, passing from the empirical rules of Egyptian land-surveying to the general propositions by which those rules were found to be justifiable, and thence to Euclid's axioms and postulates, were engaged in mathematical philosophy, according to the above definition; but when once the axioms and postulates had been reached, their deductive employment, as we find it in Euclid, belonged to mathematics in the ordinary sense. The distinction between mathematics and mathematical philosophy is one which depends upon the interest inspiring the research, and upon the stage which the research has reached; not upon the propositions with which the research is concerned.

We may state the same distinction in another way. The most obvious and easy things in mathematics are not those that come logically at the beginning; they are things that, from the point of view of logical deduction, come somewhere in the middle. Just as the easiest bodies to see are those that are neither very near nor very far, neither very small nor very great, so the easiest conceptions to grasp are those that are neither very complex nor very simple (using "simple" in a *logical* sense). And as we need two sorts of instruments, the telescope and the microscope, for the enlargement of our visual powers, so we need two sorts of instruments for the enlargement of our logical powers, one to take us forward to the higher mathematics, the other to take us backward to the logical foundations of the things that we are inclined to take for granted in mathematics. We shall find that by analysing our ordinary mathematical notions we acquire fresh insight, new powers, and the means of reaching whole new mathematical subjects by adopting fresh lines of advance after our backward journey. It is the purpose of this book to explain mathematical philosophy simply and untechnically, without enlarging upon those portions which are so doubtful or difficult that an elementary treatment is scarcely possible. A full treatment will be found in *Principia Mathematica*; [1] the treatment in the present volume is intended merely as an introduction.

[1] Cambridge University Press, vol. *i*., 1910; vol. ii., 1912; vol. iii., 1913. By Whitehead and Russell.

To the average educated person of the present day, the obvious starting-point of mathematics would be the series of whole numbers,

1, 2, 3, 4, ... etc.

Probably only a person with some mathematical knowledge would think of beginning with 0 instead of with 1, but we will presume this degree of knowledge; we will take as our starting-point the series:

$0, 1, 2, 3, \dots n, n+1, \dots$

and it is this series that we shall mean when we speak of the "series of natural numbers."

It is only at a high stage of civilization that we could take this series as our starting-point. It must have required many ages to discover that a brace of pheasants and a couple of days were both instances of the number 2: the degree of abstraction involved is far from easy. And the discovery that 1 is a number must have been difficult. As for 0, it is a very recent addition; the Greeks and Romans had no such digit. If we had been embarking upon mathematical philosophy in earlier days, we should have had to start with something less abstract than the series of natural numbers, which we should reach as a stage on our backward journey. When the logical foundations of mathematics have grown more familiar, we shall be able to start further back, at what is now a late stage in our analysis. But for the moment the natural numbers seem to represent what is easiest and most familiar in mathematics.

But though familiar, they are not understood. Very few people are prepared with a definition of what is meant by "number," or "0," or "1." It is not very difficult to see that, starting from 0, any other of the natural numbers can be reached by repeated additions of 1, but we shall have to define what we mean by "adding 1," and what we mean by "repeated." These questions are by no means easy. It was believed until recently that some, at least, of these first notions of arithmetic must be accepted as too simple and primitive to be defined. Since all terms that are defined are defined by means of other terms, it is clear that human knowledge must always be content to accept some terms as intelligible without definition, in order to have a starting-point for its definitions. It is not clear that there must be terms which are *incapable* of definition: it is possible that, however far back we go in defining, we always *might* go further still. On the other hand, it is also possible that, when analysis has been pushed far enough, we can reach terms that really are simple, and therefore logically incapable of the sort of definition that consists in analysing. This is a question which it is not necessary for us to decide; for our purposes it is sufficient to observe that, since human powers are finite, the definitions known to us must always begin somewhere, with terms undefined for the moment, though perhaps not permanently.

All traditional pure mathematics, including analytical geometry, may be regarded as consisting wholly of propositions about the natural numbers. That is to say, the terms which occur can be defined by means of the natural numbers, and the propositions can be deduced from the properties of the natural numbers—with the addition, in each case, of the ideas and propositions of pure logic.

That all traditional pure mathematics can be derived from the natural numbers is a fairly recent discovery, though it had long been suspected. Pythagoras, who believed that not only mathematics, but everything else could be deduced from numbers, was the discoverer of the most serious obstacle in the way of what is called the "arithmetising" of mathematics. It was Pythagoras who discovered the existence of incommensurables, and, in particular, the incommensurability of the side of a square and the diagonal. If the length of the side is 1 inch, the number of inches in the diagonal is the square root of 2, which appeared not to be a number at all. The problem thus raised was solved only in our own day, and was only solved *completely* by the help of the reduction of arithmetic to logic, which will be explained in following chapters. For the present, we shall take for granted the arithmetisation of mathematics, though this was a feat of the very greatest importance.

Having reduced all traditional pure mathematics to the theory of the natural numbers, the next step in logical analysis was to reduce this theory itself to the smallest set of premises and undefined terms from which it could be derived. This work was accomplished by Peano. He showed that the entire theory of the natural numbers could be derived from three primitive ideas and five primitive propositions in addition to those of pure logic. These three ideas and five propositions thus became, as it were, hostages for the whole of traditional pure mathematics. If they could be defined and proved in terms of others, so could all pure mathematics. Their logical "weight," if one may use such an expression, is equal to that of the whole series of sciences that have been deduced from the theory of the natural numbers; the truth of this whole series is assured if the truth of the five primitive propositions is guaranteed, provided, of course, that there is nothing erroneous in the purely logical apparatus which is also involved. The work of analysing mathematics is extraordinarily facilitated by this work of Peano'*s*.

The three primitive ideas in Peano'*s* arithmetic are:

0, number, successor.

By "successor" he means the next number in the natural order. That is to say, the successor of 0 is 1, the successor of 1 is 2, and so on. By "number" he means, in this connection, the class of the natural numbers. [2] He is not assuming that we know all the members of this class, but only that we know what we mean when we say that this or that is a number, just as we know what we mean when we say "Jones is a man," though we do not know all men individually.

[2] We shall use "number" in this sense in the present chapter. Afterwards the word will be used in a more general sense.

The five primitive propositions which Peano assumes are:

(1) 0 is a number.

(2) The successor of any number is a number.

(3) No two numbers have the same successor.

(4) 0 is not the successor of any number.

(5) Any property which belongs to 0, and also to the successor of every number which has the property, belongs to all numbers.

The last of these is the principle of mathematical induction. We shall have much to say concerning mathematical induction in the sequel; for the present, we are concerned with it only as it occurs in Peano's analysis of arithmetic.

Let us consider briefly the kind of way in which the theory of the natural numbers results from these three ideas and five propositions. To begin with, we define 1 as "the successor of 0," 2 as "the successor of 1," and so on. We can obviously go on as long as we like with these definitions, since, in virtue of (2), every number that we reach will have a successor, and, in virtue of (3), this cannot be any of the numbers already defined, because, if it were, two different numbers would have the same successor; and in virtue of (4) none of the numbers we reach in the series of successors can be 0. Thus the series of successors gives us an endless series of continually new numbers. In virtue of (5) all numbers come in this series, which begins with 0 and travels on through successive successors: for (*a*) 0 belongs to this series, and (*b*) if a number n belongs to it, so does its successor, whence, by mathematical induction, every number belongs to the series.

Suppose we wish to define the sum of two numbers. Taking any number m, we define $m+0$ as m, and $m+(n+1)$ as the successor of $m+n$. In virtue of (5) this gives a definition of the sum of m and n, whatever number n may be. Similarly we can define the product of any two numbers. The reader can easily convince himself that any ordinary elementary proposition of arithmetic can be proved by means of our five premises, and if he has any difficulty he can find the proof in Peano.

It is time now to turn to the considerations which make it necessary to advance beyond the standpoint of Peano, who represents the last perfection of the "arithmetisation" of mathematics, to that of Frege, who first succeeded in "logicising" mathematics, *i.e.* in reducing to logic the arithmetical notions which his predecessors had shown to be sufficient for mathematics. We shall not, in this chapter, actually give Frege's definition of number and of particular numbers, but we shall give some of the reasons why Peano's treatment is less final than it appears to be.

In the first place, Peano's three primitive ideas—namely, "0," "number," and "successor"—are capable of an infinite number of different interpretations, all of which will satisfy the five primitive propositions. We will give some examples.

(1) Let "0" be taken to mean 100, and let "number" be taken to mean the numbers from 100 onward in the series of natural numbers. Then all our primitive propositions are satisfied, even the fourth, for, though 100 is the successor of 99, 99 is not a "number" in the sense which we are now giving to the word "number." It is obvious that any number may be substituted for 100 in this example.

(2) Let "0" have its usual meaning, but let "number" mean what we usually call "even numbers," and let the "successor" of a number be what results from adding two to it. Then "1" will stand for the number two, "2" will stand for the number four, and so on; the series of "numbers" now will be

0, two, four, six, eight ...

All Peano's five premises are satisfied still.

(3) Let "0" mean the number one, let "number" mean the set

$1, \frac{1}{2}, \frac{1}{4}, \frac{1}{8}, \frac{1}{16}, \ldots$

and let "successor" mean "half." Then all Peano's five axioms will be true of this set.

It is clear that such examples might be multiplied indefinitely. In fact, given any series

$x_0, x_1, x_2, x_3, \ldots x_n, \ldots$

which is endless, contains no repetitions, has a beginning, and has no terms that cannot be reached from the beginning in a finite number of steps, we have a set of terms verifying Peano's axioms. This is easily seen, though the formal proof is somewhat long. Let "0" mean x_0, let "number" mean the whole set of terms, and let the "successor" of x_n mean x_n+1. Then

(1) "0 is a number," *i.e.* x_0 is a member of the set.

(2) "The successor of any number is a number," *i.e.* taking any term x_n in the set, x_n+1 is also in the set.

(3) "No two numbers have the same successor," *i.e.* if x_m and x_n are two different members of the set, x_m+1 and x_n+1 are different; this results from the fact that (by hypothesis) there are no repetitions in the set.

(4) "0 is not the successor of any number," *i.e.* no term in the set comes before x_0.

(5) This becomes: Any property which belongs to x_0, and belongs to x_n+1 provided it belongs to x_n, belongs to all the *x's*.

This follows from the corresponding property for numbers.

A series of the form

$x_0, x_1, x_2, \ldots x_n, \ldots$

in which there is a first term, a successor to each term (so that there is no last term), no repetitions, and every term can be reached from the start in a finite number of steps, is called a *progression.* Progressions are of great importance in the principles of mathematics. As we have just seen, every progression verifies Peano's five axioms. It can be proved, conversely, that every series which verifies Peano's five axioms is a progression. Hence these five axioms may be used to define the class of progressions: "progressions" are "those series which verify these five axioms." Any progression may be taken as the basis of pure mathematics: we may give the name "0" to its first term, the name "number" to the whole set of its terms, and the name "successor" to the next in the progression. The progression need not be composed of numbers: it may be composed of points in space, or moments of time, or any other terms of which there is an infinite supply. Each different progression will give rise to a different interpretation of all the propositions of traditional pure mathematics; all these possible interpretations will be equally true.

In Peano's system there is nothing to enable us to distinguish between these different interpretations of his primitive ideas. It is assumed that we know what is meant by "0," and that we shall not suppose that this symbol means 100 or Cleopatra's Needle or any of the other things that it might mean.

This point, that "0" and "number" and "successor" cannot be defined by means of Peano's five axioms, but must be independently understood, is important. We want our numbers not merely to verify mathematical formulæ, but to apply in the right way to common objects. We want to have ten fingers and two eyes and one nose. A system in which "1" meant 100, and "2" meant 101, and so on, might be all right for pure mathematics, but would not suit daily life. We want "0" and "number" and "successor" to have meanings which will give us the right allowance of fingers and eyes and noses. We have already some knowledge (though not sufficiently articulate or analytic) of what we mean by "1" and "2" and so on, and our use of numbers in arithmetic must conform to this knowledge. We cannot secure that this shall be the case by Peano's method; all that we can do, if we adopt his method, is to say "we know what we mean by '0' and 'number' and 'successor,' though we cannot explain what we mean in terms of other simpler concepts." It is quite legitimate to say this when we must, and at *some* point we all must; but it is the object of mathematical philosophy to put off saying it as long as possible. By the logical theory of arithmetic we are able to put it off for a very long time.

It might be suggested that, instead of setting up "0" and "number" and "successor" as terms of which we know the meaning although we cannot define them, we might let them stand for *any* three terms that verify Peano's five axioms. They will then no longer be terms which have a meaning that is definite though undefined: they will be "variables," terms concerning which we make certain hypotheses, namely, those stated in the five axioms, but which are otherwise undetermined. If we adopt this plan, our theorems will not be proved concerning an ascertained set of terms called "the natural numbers," but concerning all sets of terms having certain properties. Such a procedure is not fallacious; indeed for certain purposes it represents a valuable generalization. But from two points of view it fails to give an adequate basis for arithmetic. In the first place, it does not enable us to know whether there are any sets of terms verifying Peano's axioms; it does not even give the faintest suggestion of any way of discovering whether there are such sets. In the second place, as already observed, we want our numbers to be such as can be used for counting common objects, and this requires that our numbers should have a *definite* meaning, not merely that they should have certain formal properties. This definite meaning is defined by the logical theory of arithmetic.

CHAPTER II

DEFINITION OF NUMBER

The question "What is a number?" is one which has been often asked, but has only been correctly answered in our own time. The answer was given by Frege in 1884, in his *Grundlagen der Arithmetik*. [3] Although this book is quite short, not difficult, and of the very highest importance, it attracted almost no attention, and the definition of number which it contains remained practically unknown until it was rediscovered by the present author in 1901.

[3] The same answer is given more fully and with more development in his *Grundgesetze der Arithmetik*, vol. *i.*, 1893.

In seeking a definition of number, the first thing to be clear about is what we may call the grammar of our inquiry. Many philosophers, when attempting to define number, are really setting to work to define plurality, which is quite a different thing. *Number* is what is characteristic of numbers, as *man* is what is characteristic of men. A plurality is not an instance of number, but of some particular number. A trio of men, for example, is an instance of the number 3, and the number 3 is an instance of number; but the trio is not an instance of number. This point may seem elementary and scarcely worth mentioning; yet it has proved too subtle for the philosophers, with few exceptions.

A particular number is not identical with any collection of terms having that number: the number 3 is not identical with the trio consisting of Brown, Jones, and Robinson. The number 3 is something which all trios have in common, and which distinguishes them from other collections. A number is something that characterizes certain collections, namely, those that have that number.

Instead of speaking of a "collection," we shall as a rule speak of a "class," or sometimes a "set." Other words used in mathematics for the same thing are "aggregate" and "manifold." We shall have much to say later on about classes. For the present, we will say as little as possible. But there are some remarks that must be made immediately.

A class or collection may be defined in two ways that at first sight seem quite distinct. We may enumerate its members, as when we say, "The collection I mean is Brown, Jones, and Robinson." Or we may mention a defining property, as when we speak of "mankind" or "the inhabitants of London." The definition which enumerates is called a definition by "extension," and the one which mentions a defining property is called a definition by "intension." Of these two kinds of definition, the one by intension is logically more fundamental. This is shown by two considerations: (1) that the extensional definition can always be reduced to an intensional one; (2) that the intensional one often cannot even theoretically be reduced to the extensional one. Each of these points needs a word of explanation.

(1) Brown, Jones, and Robinson all of them possess a certain property which is possessed by nothing else in the whole universe, namely, the property of being either Brown or Jones or Robinson. This property can be used to give a definition by intension of the class consisting of Brown and Jones and Robinson. Consider such a formula as "x is Brown or x is Jones or x is Robinson." This formula will be true for just three x's, namely, Brown and Jones and Robinson. In this respect it resembles a cubic equation with its three roots. It may be taken as assigning a property common to the members of the class consisting of these three men, and peculiar to them. A similar treatment can obviously be applied to any other class given in extension.

(2) It is obvious that in practice we can often know a great deal about a class without being able to enumerate its members. No one man could actually enumerate all men, or even all the inhabitants of London, yet a great deal is known about each of these classes. This is enough to show that definition by extension is not *necessary* to knowledge about a class. But when we come to consider infinite classes, we find that enumeration is not even theoretically possible for beings who only live for a finite time. We cannot enumerate all the natural numbers: they are 0, 1, 2, 3, *and so on*. At some point we must content ourselves with "and so on." We cannot enumerate all fractions or all irrational numbers, or all of any other infinite collection. Thus our knowledge in regard to all such collections can only be derived from a definition by intension.

These remarks are relevant, when we are seeking the definition of number, in three different ways. In the first place, numbers themselves form an infinite collection, and cannot therefore be defined by enumeration. In the second place, the collections having a given number of terms themselves presumably form an infinite collection: it is to be presumed, for example, that there are an infinite collection of trios in the world, for if this were not the case the total number of things in the world would be finite, which, though possible, seems unlikely. In the third place, we wish to define "number" in such a way that infinite numbers may be possible; thus we must be able to speak of the number of terms in an infinite collection, and such a collection must be defined by intension, *i.e.* by a property common to all its members and peculiar to them.

For many purposes, a class and a defining characteristic of it are practically interchangeable. The vital difference between the two consists in the fact that there is only one class having a given set of members, whereas there are always many different characteristics by which a given class may be defined. Men may be defined as featherless bipeds, or as rational animals, or (more correctly) by the traits by which Swift delineates the Yahoos. It is this fact that a defining characteristic is never unique which makes classes useful; otherwise we could be content with the properties common and peculiar to their members. [4] Any one of these properties can be used in place of the class whenever uniqueness is not important.

[4] As will be explained later, classes may be regarded as logical fictions, manufactured out of defining characteristics. But for the present it will simplify our exposition to treat classes as if they were real.

Returning now to the definition of number, it is clear that number is a way of bringing together certain collections, namely, those that have a given number of terms. We can suppose all couples in one bundle, all trios in another, and so on. In this way we obtain various bundles of collections, each bundle consisting of all the collections that have a certain number of terms. Each bundle is a class whose members are collections, *i.e.* classes; thus each is a class of classes. The bundle consisting of all couples, for example, is a class of classes: each couple is a class with two members, and the whole bundle of couples is a class with an infinite number of members, each of which is a class of two members.

How shall we decide whether two collections are to belong to the same bundle? The answer that suggests itself is: "Find out how many members each has, and put them in the same bundle if they have the same number of members." But this presupposes that we have defined numbers, and that we know how to discover how many terms a collection has. We are so used to the operation of counting that such a presupposition might easily pass unnoticed. In fact, however, counting, though familiar, is logically a very complex operation; moreover it is only available, as a means of discovering how many terms a collection has, when the collection is finite. Our definition of number must not assume in advance that all numbers are finite; and we cannot in any case, without a vicious circle, use counting to define numbers, because numbers are used in counting. We need, therefore, some other method of deciding when two collections have the same number of terms.

In actual fact, it is simpler logically to find out whether two collections have the same number of terms than it is to define what that number is. An illustration will make this clear. If there were no polygamy or polyandry anywhere in the world, it is clear that the number of husbands living at any moment would be exactly the same as the number of wives. We do not need a census to assure us of this, nor do we need to know what is the actual number of husbands and of wives. We know the number must be the same in both collections, because each husband has one wife and each wife has one husband. The relation of husband and wife is what is called "one-one."

A relation is said to be "one-one" when, if x has the relation in question to y, no other term x' has the same relation to y, and x does not have the same relation to any term y' other than y. When only the first of these two conditions is fulfilled, the relation is called "one-many"; when only the second is fulfilled, it is called "many-one." It should be observed that the number 1 is not used in these definitions.

In Christian countries, the relation of husband to wife is one-one; in Mahometan countries it is one-many; in Tibet it is many-one. The relation of father to son is one-many; that of son to father is many-one, but that of eldest son to father is one-one. If n is any number, the relation of n to $n+1$ is one-one; so is the relation of n to $2n$ or to $3n$. When we are considering only positive numbers, the relation of n to n^2 is one-one; but when negative numbers are admitted, it becomes two-one, since n and $-n$ have the same square. These instances should suffice to make clear the notions of one-one, one-many, and many-one relations, which play a great part in the principles of mathematics, not only in relation to the definition of numbers, but in many other connections.

Two classes are said to be "similar" when there is a one-one relation which correlates the terms of the one class each with one term of the other class, in the same manner in which the relation of marriage correlates husbands with wives. A few preliminary definitions will help us to state this definition more precisely. The class of those terms that have a given relation to something or other is called the *domain* of that relation: thus fathers are the domain of the relation of father to child, husbands are the domain of the relation of husband to wife, wives are the domain of the relation of wife to husband, and husbands and wives together are the domain of the relation of marriage. The relation of wife to husband is called the *converse* of the relation of husband to wife. Similarly *less* is the converse of *greater, later* is the converse of *earlier*, and so on. Generally, the converse of a given relation is that relation which holds between y and x whenever the given relation holds between x and y. The *converse domain* of a relation is the domain of its converse: thus the class of wives is the converse domain of the relation of husband to wife. We may now state our definition of similarity as follows:—

One class is said to be "similar" to another when there is a one-one relation of which the one class is the domain, while the other is the converse domain.

It is easy to prove (1) that every class is similar to itself, (2) that if a class α is similar to a class β, then β is similar to α, (3) that if α is similar to β and β to γ, then α is similar to γ. A relation is said to be *reflexive* when it possesses the first of these properties, *symmetrical* when it possesses the second, and *transitive* when it possesses the third. It is obvious that a relation which is symmetrical and transitive must be reflexive throughout its domain. Relations which possess these properties are an important kind, and it is worthwhile to note that similarity is one of this kind of relations.

It is obvious to common sense that two finite classes have the same number of terms if they are similar, but not otherwise. The act of counting consists in establishing a one-one correlation between the set of objects counted and the natural numbers (excluding 0) that are used up in the process. Accordingly common sense concludes that there are as many objects in the set to be counted as there are numbers up to the last number used in the counting. And we also know that, so long as we confine ourselves to finite numbers, there are just n numbers from 1 up to n. Hence it follows that the last number used in counting a collection is the number of terms in the collection, provided the collection is finite. But this result, besides being only applicable to finite collections, depends upon and assumes the fact that two classes which are similar have the same number of terms; for what we do when we count (say) 10 objects is to show that the set of these objects is similar to the set of numbers 1 to 10. The notion of similarity is logically presupposed in the operation of counting, and is logically simpler though less familiar. In counting, it is necessary to take the objects counted in a certain order, as first, second, third, etc., but order is not of the essence of number: it is an irrelevant addition, an unnecessary complication from the logical point of view. The notion of similarity does not demand an order: for example, we saw that the number of husbands is the same as the number of wives, without having to establish an order of precedence among them. The notion of similarity also does not require that the classes which are similar should be finite. Take, for example, the natural numbers (excluding 0) on the one hand, and the fractions which have 1 for their numerator on the other hand: it is obvious that we can correlate 2 with ½, 3 with ⅓, and so on, thus proving that the two classes are similar.

We may thus use the notion of "similarity" to decide when two collections are to belong to the same bundle, in the sense in which we were asking this question earlier in this chapter. We want to make one bundle containing the class that has no members: this will be for the number 0. Then we want a bundle of all the classes that have one member: this will be for the number 1. Then, for the number 2, we want a bundle consisting of all couples; then one of all trios; and so on. Given any collection, we can define the bundle it is to belong to as being the class of all those collections that are "similar" to it. It is very easy to see that if (for example) a collection has three members, the class of all those collections that are similar to it will be the class of trios. And whatever number of terms a collection may have, those collections that are "similar" to it will have the same number of terms. We may take this as a *definition* of "having the same number of terms." It is obvious that it gives results conformable to usage so long as we confine ourselves to finite collections.

So far we have not suggested anything in the slightest degree paradoxical. But when we come to the actual definition of numbers we cannot avoid what must at first sight seem a paradox, though this impression will soon wear off. We naturally think that the class of couples (for example) is something different from the number 2. But there is no doubt about the class of couples: it is indubitable and not difficult to define, whereas the number 2, in any other sense, is a metaphysical entity about which we can never feel sure that it exists or that we have tracked it down. It is therefore more prudent to content ourselves with the class of couples, which we are sure of, than to hunt for a problematical number 2 which must always remain elusive. Accordingly we set up the following definition:—

The number of a class is the class of all those classes that are similar to it.

Thus the number of a couple will be the class of all couples. In fact, the class of all couples will be the number 2, according to our definition. At the expense of a little oddity, this definition secures definiteness and indubitableness; and it is not difficult to prove that numbers so defined have all the properties that we expect numbers to have.

We may now go on to define numbers in general as any one of the bundles into which similarity collects classes. A number will be a set of classes such as that any two are similar to each other, and none outside the set are similar to any inside the set. In other words, a number (in general) is any collection which is the number of one of its members; or, more simply still:

A number is anything which is the number of some class.

Such a definition has a verbal appearance of being circular, but in fact it is not. We define "the number of a given class" without using the notion of number in general; therefore we may define number in general in terms of "the number of a given class" without committing any logical error.

Definitions of this sort are in fact very common. The class of fathers, for example, would have to be defined by first defining what it is to be the father of somebody; then the class of fathers will be all those who are somebody'*s* father. Similarly if we want to define square numbers (say), we must first define what we mean by saying that one number is the square of another, and then define square numbers as those that are the squares of other numbers. This kind of procedure is very common, and it is important to realize that it is legitimate and even often necessary.

We have now given a definition of numbers which will serve for finite collections. It remains to be seen how it will serve for infinite collections. But first we must decide what we mean by "finite" and "infinite," which cannot be done within the limits of the present chapter.

CHAPTER III

FINITUDE AND MATHEMATICAL INDUCTION

The series of natural numbers, as we saw in Chapter I., can all be defined if we know what we mean by the three terms "0," "number," and "successor." But we may go a step farther: we can define all the natural numbers if we know what we mean by "0" and "successor." It will help us to understand the difference between finite and infinite to see how this can be done, and why the method by which it is done cannot be extended beyond the finite. We will not yet consider how "0" and "successor" are to be defined: we will for the moment assume that we know what these terms mean, and show how thence all other natural numbers can be obtained.

It is easy to see that we can reach any assigned number, say 30,000. We first define "1" as "the successor of 0," then we define "2" as "the successor of 1," and so on. In the case of an assigned number, such as 30,000, the proof that we can reach it by proceeding step by step in this fashion may be made, if we have the patience, by actual experiment: we can go on until we actually arrive at 30,000. But although the method of experiment is available for each particular natural number, it is not available for proving the general proposition that *all* such numbers can be reached in this way, *i.e.* by proceeding from 0 step by step from each number to its successor. Is there any other way by which this can be proved?

Let us consider the question the other way round. What are the numbers that can be reached, given the terms "0" and "successor"? Is there any way by which we can define the whole class of such numbers? We reach 1, as the successor of 0; 2, as the successor of 1; 3, as the successor of 2; and so on. It is this "and so on" that we wish to replace by something less vague and indefinite. We might be tempted to say that "and so on" means that the process of proceeding to the successor may be repeated *any finite number* of times; but the problem upon which we are engaged is the problem of defining "finite number," and therefore we must not use this notion in our definition. Our definition must not assume that we know what a finite number is.

The key to our problem lies in *mathematical induction*. It will be remembered that, in Chapter I., this was the fifth of the five primitive propositions which we laid down about the natural numbers. It stated that any property which belongs to 0, and to the successor of any number which has the property, belongs to all the natural numbers. This was then presented as a principle, but we shall now adopt it as a definition. It is not difficult to see that the terms obeying it are the same as the numbers that can be reached from 0 by successive steps from next to next, but as the point is important we will set forth the matter in some detail.

We shall do well to begin with some definitions, which will be useful in other connections also.

A property is said to be "hereditary" in the natural-number series if, whenever it belongs to a number n, it also belongs to $n+1$, the successor of n. Similarly a class is said to be "hereditary" if, whenever n is a member of the class, so is $n+1$. It is easy to *see*, though we are not yet supposed to know, that to say a property is hereditary is equivalent to saying that it belongs to all the natural numbers not less than some one of them, *e.g.* it must belong to all that are not less than 100, or all that are not less than 1000, or it may be that it belongs to all that are not less than 0, *i.e.* to all without exception.

A property is said to be "inductive" when it is a hereditary property which belongs to 0. Similarly a class is "inductive" when it is a hereditary class of which 0 is a member.

Given a hereditary class of which 0 is a member, it follows that 1 is a member of it, because a hereditary class contains the successors of its members, and 1 is the successor of 0. Similarly, given a hereditary class of which 1 is a member, it follows that 2 is a member of it; and so on. Thus we can prove by a step-by-step procedure that any assigned natural number, say 30,000, is a member of every inductive class.

We will define the "posterity" of a given natural number with respect to the relation "immediate predecessor" (which is the converse of "successor") as all those terms that belong to every hereditary class to which the given number belongs. It is again easy to *see* that the posterity of a natural number consists of itself and all greater natural numbers; but this also we do not yet officially know.

By the above definitions, the posterity of 0 will consist of those terms which belong to every inductive class.

It is now not difficult to make it obvious that the posterity of 0 is the same set as those terms that can be reached from 0 by successive steps from next to next. For, in the first place, 0 belongs to both these sets (in the sense in which we have defined our terms); in the second place, if n belongs to both sets, so does $n+1$. It is to be observed that we are dealing here with the kind of matter that does not admit of precise proof, namely, the comparison of a relatively vague idea with a relatively precise one. The notion of "those terms that can be reached from 0 by successive steps from next to next" is vague, though it *seems* as if it conveyed a definite meaning; on the other hand, "the posterity of 0" is precise and explicit just where the other idea is hazy. It may be taken as giving what we *meant* to mean when we spoke of the terms that can be reached from 0 by successive steps.

We now lay down the following definition:—

The "natural numbers" are the posterity of 0 with respect to the relation "immediate predecessor" (which is the converse of "successor").

We have thus arrived at a definition of one of Peano's three primitive ideas in terms of the other two. As a result of this definition, two of his primitive propositions—namely, the one asserting that 0 is a number and the one asserting mathematical induction—become unnecessary, since they result from the definition. The one asserting that the successor of a natural number is a natural number is only needed in the weakened form "every natural number has a successor."

We can, of course, easily define "0" and "successor" by means of the definition of number in general which we arrived at in Chapter II. The number 0 is the number of terms in a class which has no members, *i.e.* in the class which is called the "null-class." By the general definition of number, the number of terms in the null-class is the set of all classes similar to the null-class, *i.e.* (as is easily proved) the set consisting of the null-class all alone, *i.e.* the class whose only member is the null-class. (This is not identical with the null-class: it has one member, namely, the null-class, whereas the null-class itself has no members. A class which has one member is never identical with that one member, as we shall explain when we come to the theory of classes.) Thus we have the following purely logical definition:—

0 is the class whose only member is the null-class.

It remains to define "successor." Given any number n, let α be a class which has n members, and let x be a term which is not a member of α. Then the class consisting of α with x added on will have $n+1$ members. Thus we have the following definition:—

The successor of the number of terms in the class α *is the number of terms in the class consisting of* α *together with x, where x is any term not belonging to the class.*

Certain niceties are required to make this definition perfect, but they need not concern us. [5] It will be remembered that we have already given (in Chapter II.) a logical definition of the number of terms in a class, namely, we defined it as the set of all classes that are similar to the given class.

[5] See *Principia Mathematica*, vol. ii. *110.

We have thus reduced Peano's three primitive ideas to ideas of logic: we have given definitions of them which make them definite, no longer capable of an infinity of different meanings, as they were when they were only determinate to the extent of obeying Peano's five axioms. We have removed them from the fundamental apparatus of terms that must be merely apprehended, and have thus increased the deductive articulation of mathematics.

As regards the five primitive propositions, we have already succeeded in making two of them demonstrable by our definition of "natural number." How stands it with the remaining three? It is very easy to prove that 0 is not the successor of any number, and that the successor of any number is a number. But there is a difficulty about the remaining primitive proposition, namely, "no two numbers have the same successor." The difficulty does not arise unless the total number of individuals in the universe is finite; for given two numbers m and n, neither of which is the total number of individuals in the universe, it is easy to prove that we cannot have $m+1=n+1$ unless we have $m=n$. But let us suppose that the total number of individuals in the universe were (say) 10; then there would be no class of 11 individuals, and the number 11 would be the null-class. So would the number 12. Thus we should have 11=12; therefore the successor of 10 would be the same as the successor of 11, although 10 would not be the same as 11. Thus we should have two different numbers with the same successor. This failure of the third axiom cannot arise, however, if the number of individuals in the world is not finite. We shall return to this topic at a later stage. [6]

[6] See Chapter XIII.

Assuming that the number of individuals in the universe is not finite, we have now succeeded not only in defining Peano's three primitive ideas, but in seeing how to prove his five primitive propositions, by means of primitive ideas and propositions belonging to logic. It follows that all pure mathematics, in so far as it is deducible from the theory of the natural numbers, is only a prolongation of logic. The extension of this result to those modern branches of mathematics which are not deducible from the theory of the natural numbers offers no difficulty of principle, as we have shown elsewhere. [7]

[7]. For geometry, in so far as it is not purely analytical, see *Principles of Mathematics*, part vi.; for rational dynamics, *ibid.*, part vii.

The process of mathematical induction, by means of which we defined the natural numbers, is capable of generalization. We defined the natural numbers as the "posterity" of 0 with respect to the relation of a number to its immediate successor. If we call this relation N, any number m will have this relation to $m+1$. A property is "hereditary with respect to N," or simply "N-hereditary," if, whenever the property belongs to a number m, it also belongs to $m+1$, *i.e.* to the number to which m has the relation N. And a number n will be said to belong to the "posterity" of m with respect to the relation N if n has every N-hereditary property belonging to m. These definitions can all be applied to any other relation just as well as to N. Thus if R is any relation whatever, we can lay down the following definitions: [8]—

[8] These definitions, and the generalized theory of induction, are due to Frege, and were published so long ago as 1879 in his *Begriffsschrift*. In spite of the great value of this work, I was, I believe, the first person who ever read it—more than twenty years after its publication.

A property is called "R-hereditary" when, if it belongs to a term x, and x has the relation R to y, then it belongs to y.

A class is R-hereditary when its defining property is R-hereditary.

A term x is said to be an "R-ancestor" of the term y if y has every R-hereditary property that x has, provided x is a term which has the relation R to something or to which something has the relation R. (This is only to exclude trivial cases.)

The "R-posterity" of x is all the terms of which x is an R-ancestor.

We have framed the above definitions so that if a term is the ancestor of anything it is its own ancestor and belongs to its own posterity. This is merely for convenience.

It will be observed that if we take for R the relation "parent," "ancestor" and "posterity" will have the usual meanings, except that a person will be included among his own ancestors and posterity. It is, of course, obvious at once that "ancestor" must be capable of definition in terms of "parent," but until Frege developed his generalised theory of induction, no one could have defined "ancestor" precisely in terms of "parent." A brief consideration of this point will serve to show the importance of the theory. A person confronted for the first time with the problem of defining "ancestor" in terms of "parent" would naturally say that A is an ancestor of Z if, between A and Z, there are a certain number of people, B, C, ..., of whom B is a child of A, each is a parent of the next, until the last, who is a parent of Z. But this definition is not adequate unless we add that the number of intermediate terms is to be finite. Take, for example, such a series as the following:—

$$-1, -\frac{1}{2}, -\frac{1}{4}, -\frac{1}{8}, \ldots \frac{1}{8}, \frac{1}{4}, \frac{1}{2}, 1.$$

Here we have first a series of negative fractions with no end, and then a series of positive fractions with no beginning. Shall we say that, in this series, −1/8 is an ancestor of 1/8? It will be so according to the beginner's definition suggested above, but it will not be so according to any definition which will give the kind of idea that we wish to define. For this purpose, it is essential that the number of intermediaries should be finite. But, as we saw, "finite" is to be defined by means of mathematical induction, and it is simpler to define the ancestral relation generally at once than to define it first only for the case of the relation of n to $n+1$, and then extend it to other cases. Here, as constantly elsewhere, generality from the first, though it may require more thought at the start, will be found in the long run to economize thought and increase logical power.

The use of mathematical induction in demonstrations was, in the past, something of a mystery. There seemed no reasonable doubt that it was a valid method of proof, but no one quite knew why it was valid. Some believed it to be really a case of induction, in the sense in which that word is used in logic. Poincaré [9] considered it to be a principle of the utmost importance, by means of which an infinite number of syllogisms could be condensed into one argument. We now know that all such views are mistaken, and that mathematical induction is a definition, not a principle. There are some numbers to which it can be applied, and there are others (as we shall see in Chapter VIII.) to which it cannot be applied. We *define* the "natural numbers" as those to which proofs by mathematical induction can be applied, *i.e.* as those that possess all inductive properties. It follows that such proofs can be applied to the natural numbers, not in virtue of any mysterious intuition or axiom or principle, but as a purely verbal proposition. If "quadrupeds" are defined as animals having four legs, it will follow that animals that have four legs are quadrupeds; and the case of numbers that obey mathematical induction is exactly similar.

[9] *Science and Method*, chap. iv.

We shall use the phrase "inductive numbers" to mean the same set as we have hitherto spoken of as the "natural numbers." The phrase "inductive numbers" is preferable as affording a reminder that the definition of this set of numbers is obtained from mathematical induction.

Mathematical induction affords, more than anything else, the essential characteristic by which the finite is distinguished from the infinite. The principle of mathematical induction might be stated popularly in some such form as "what can be inferred from next to next can be inferred from first to last." This is true when the number of intermediate steps between first and last is finite, not otherwise. Anyone who has ever watched a goods train beginning to move will have noticed how the impulse is communicated with a jerk from each truck to the next, until at last even the hindmost truck is in motion. When the train is very long, it is a very long time before the last truck moves. If the train were infinitely long, there would be an infinite succession of jerks, and the time would never come when the whole train would be in motion. Nevertheless, if there were a series of trucks no longer than the series of inductive numbers (which, as we shall see, is an instance of the smallest of infinites), every truck would begin to move sooner or later if the engine persevered, though there would always be other trucks further back which had not yet begun to move. This image will help to elucidate the argument from next to next, and its connection with finitude. When we come to infinite numbers, where arguments from mathematical induction will be no longer valid, the properties of such numbers will help to make clear, by contrast, the almost unconscious use that is made of mathematical induction where finite numbers are concerned.

CHAPTER IV

THE DEFINITION OF ORDER

We have now carried our analysis of the series of natural numbers to the point where we have obtained logical definitions of the members of this series, of the whole class of its members, and of the relation of a number to its immediate successor. We must now consider the *serial* character of the natural numbers in the order 0, 1, 2, 3, ... We ordinarily think of the numbers as in this *order*, and it is an essential part of the work of analysing our data to seek a definition of "order" or "series" in logical terms.

The notion of order is one which has enormous importance in mathematics. Not only the integers, but also rational fractions and all real numbers have an order of magnitude, and this is essential to most of their mathematical properties. The order of points on a line is essential to geometry; so is the slightly more complicated order of lines through a point in a plane, or of planes through a line. Dimensions, in geometry, are a development of order. The conception of a *limit*, which underlies all higher mathematics, is a serial conception. There are parts of mathematics which do not depend upon the notion of order, but they are very few in comparison with the parts in which this notion is involved.

In seeking a definition of order, the first thing to realize is that no set of terms has just *one* order to the exclusion of others. A set of terms has all the orders of which it is capable. Sometimes one order is so much more familiar and natural to our thoughts that we are inclined to regard it as the order of that set of terms; but this is a mistake. The natural numbers—or the "inductive" numbers, as we shall also call them—occur to us most readily in order of magnitude; but they are capable of an infinite number of other arrangements. We might, for example, consider first all the odd numbers and then all the even numbers; or first 1, then all the even numbers, then all the odd multiples of 3, then all the multiples of 5 but not of 2 or 3, then all the multiples of 7 but not of 2 or 3 or 5, and so on through the whole series of primes. When we say that we "arrange" the numbers in these various orders, that is an inaccurate expression: what we really do is to turn our attention to certain relations between the natural numbers, which themselves generate such-and-such an arrangement. We can no more "arrange" the natural numbers than we can the starry heavens; but just as we may notice among the fixed stars either their order of brightness or their distribution in the sky, so there are various relations among numbers which may be observed, and which give rise to various different orders among numbers, all equally legitimate. And what is true of numbers is equally true of points on a line or of the moments of time: one order is more familiar, but others are equally valid. We might, for example, take first, on a line, all the points that have integral co-ordinates, then all those that have non-integral rational co-ordinates, then all those that have algebraic non-rational co-ordinates, and so on, through any set of complications we please. The resulting order will be one which the points of the line certainly have, whether we choose to notice it or not; the only thing that is arbitrary about the various orders of a set of terms is our attention, for the terms themselves have always all the orders of which they are capable.

One important result of this consideration is that we must not look for the definition of order in the nature of the set of terms to be ordered, since one set of terms has many orders. The order lies, not in the *class* of terms, but in a relation among the members of the class, in respect of which some appear as earlier and some as later. The fact that a class may have many orders is due to the fact that there can be many relations holding among the members of one single class. What properties must a relation have in order to give rise to an order?

The essential characteristics of a relation which is to give rise to order may be discovered by considering that in respect of such a relation we must be able to say, of any two terms in the class which is to be ordered, that one "precedes" and the other "follows." Now, in order that we may be able to use these words in the way in which we should naturally understand them, we require that the ordering relation should have three properties:—

(1) If x precedes y, y must not also precede x. This is an obvious characteristic of the kind of relations that lead to series. If x is less than y, y is not also less than x. If x is earlier in time than y, y is not also earlier than x. If x is to the left of y, y is not to the left of x. On the other hand, relations which do not give rise to series often do not have this property. If x is a brother or sister of y, y is a brother or sister of x. If x is of the same height as y, y is of the same height as x. If x is of a different height from y, y is of a different height from x. In all these cases, when the relation holds between x and y, it also holds between y and x. But with serial relations such a thing cannot happen. A relation having this first property is called *asymmetrical*.

(2) If x precedes y and y precedes z, x must precede z. This may be illustrated by the same instances as before: *less, earlier, left of*. But as instances of relations which do *not* have this property only two of our previous three instances will serve. If x is brother or sister of y, and y of z, x may not be brother or sister of z, since x and z may be the same person. The same applies to difference of height, but not to sameness of height, which has our second property but not our first. The relation "father," on the other hand, has our first property but not our second. A relation having our second property is called *transitive*.

(3) Given any two terms of the class which is to be ordered, there must be one which precedes and the other which follows. For example, of any two integers, or fractions, or real numbers, one is smaller and the other greater; but of any two complex numbers this is not true. Of any two moments in time, one must be earlier than the other; but of events, which may be simultaneous, this cannot be said. Of two points on a line, one must be to the left of the other. A relation having this third property is called *connected*.

When a relation possesses these three properties, it is of the sort to give rise to an order among the terms between which it holds; and wherever an order exists, some relation having these three properties can be found generating it.

Before illustrating this thesis, we will introduce a few definitions.

(1) A relation is said to be an aliorelative, [10] or to *be contained in* or *imply diversity*, if no term has this relation to itself. Thus, for example, "greater," "different in size," "brother," "husband," "father" are aliorelatives; but "equal," "born of the same parents," "dear friend" are not.

[10] This term is due to C. S. Peirce.

(2) The *square* of a relation is that relation which holds between two terms x and z when there is an intermediate term y such that the given relation holds between x and y and between y and z. Thus "paternal grandfather" is the square of "father," "greater by 2" is the square of "greater by 1," and so on.

(3) The *domain* of a relation consists of all those terms that have the relation to something or other, and the *converse domain* consists of all those terms to which something or other has the relation. These words have been already defined, but are recalled here for the sake of the following definition:—

(4) The *field* of a relation consists of its domain and converse domain together.

(5) One relation is said to *contain* or *be implied by* another if it holds whenever the other holds.

It will be seen that an *asymmetrical* relation is the same thing as a relation whose square is an aliorelative. It often happens that a relation is an aliorelative without being asymmetrical, though an asymmetrical relation is always an aliorelative. For example, "spouse" is an aliorelative, but is symmetrical, since if x is the spouse of y, y is the spouse of x. But among *transitive* relations, all aliorelatives are asymmetrical as well as *vice versa*.

From the definitions it will be seen that a *transitive* relation is one which is implied by its square, or, as we also say, "contains" its square. Thus "ancestor" is transitive, because an ancestor's ancestor is an ancestor; but "father" is not transitive, because a father's father is not a father. A transitive aliorelative is one which contains its square and is contained in diversity; or, what comes to the same thing, one whose square implies both it and diversity—because, when a relation is transitive, asymmetry is equivalent to being an aliorelative.

A relation is *connected* when, given any two different terms of its field, the relation holds between the first and the second or between the second and the first (not excluding the possibility that both may happen, though both cannot happen if the relation is asymmetrical).

It will be seen that the relation "ancestor," for example, is an aliorelative and transitive, but not connected; it is because it is not connected that it does not suffice to arrange the human race in a series.

The relation "less than or equal to," among numbers, is transitive and connected, but not asymmetrical or an aliorelative.

The relation "greater or less" among numbers is an aliorelative and is connected, but is not transitive, for if x is greater or less than y, and y is greater or less than z, it may happen that x and z are the same number.

Thus the three properties of being (1) an aliorelative, (2) transitive, and (3) connected, are mutually independent, since a relation may have any two without having the third.

We now lay down the following definition:—

A relation is *serial* when it is an aliorelative, transitive, and connected; or, what is equivalent, when it is asymmetrical, transitive, and connected.

A *series* is the same thing as a serial relation.

It might have been thought that a series should be the *field* of a serial relation, not the serial relation itself. But this would be an error. For example,

1, 2, 3; 1, 3, 2; 2, 3, 1; 2, 1, 3; 3, 1, 2; 3, 2, 1

are six different series which all have the same field. If the field were the series, there could only be one series with a given field. What distinguishes the above six series is simply the different ordering relations in the six cases. Given the ordering relation, the field and the order are both determinate. Thus the ordering relation may be taken to be the series, but the field cannot be so taken.

Given any serial relation, say P, we shall say that, in respect of this relation, x "precedes" y if x has the relation P to y, which we shall write "xPy" for short. The three characteristics which P must have in order to be serial are:

(1) We must never have xPx, *i.e.* no term must precede itself.
(2) P^2 must imply P, *i.e.* if x precedes y and y precedes z, x must precede z.
(3) If x and y are two different terms in the field of P, we shall have xPy or yPx, *i.e.* one of the two must precede the other.

The reader can easily convince himself that, where these three properties are found in an ordering relation, the characteristics we expect of series will also be found, and vice versa. We are therefore justified in taking the above as a definition of order or series. And it will be observed that the definition is effected in purely logical terms.

Although a transitive asymmetrical connected relation always exists wherever there is a series, it is not always the relation which would most naturally be regarded as generating the series. The natural-number series may serve as an illustration. The relation we assumed in considering the natural numbers was the relation of immediate succession, *i.e.* the relation between consecutive integers. This relation is asymmetrical, but not transitive or connected. We can, however, derive from it, by the method of mathematical induction, the "ancestral" relation which we considered in the preceding chapter. This relation will be the same as "less than or equal to" among inductive integers. For purposes of generating the series of natural numbers, we want the relation "less than," excluding "equal to." This is the relation of m to n when m is an ancestor of n but not identical with n, or (what comes to the same thing) when the successor of m is an ancestor of n in the sense in which a number is its own ancestor. That is to say, we shall lay down the following definition:—

An inductive number m is said to be *less than* another number n when n possesses every hereditary property possessed by the successor of m.

It is easy to see, and not difficult to prove, that the relation "less than," so defined, is asymmetrical, transitive, and connected, and has the inductive numbers for its field. Thus by means of this relation the inductive numbers acquire an order in the sense in which we defined the term "order," and this order is the so-called "natural" order, or order of magnitude.

The generation of series by means of relations more or less resembling that of n to $n+1$ is very common. The series of the Kings of England, for example, is generated by relations of each to his successor. This is probably the easiest way, where it is applicable, of conceiving the generation of a series. In this method we pass on from each term to the next, as long as there is a next, or back to the one before, as long as there is one before. This method always requires the generalized form of mathematical induction in order to enable us to define "earlier" and "later" in a series so generated. On the analogy of "proper fractions," let us give the name "proper posterity of x with respect to R" to the class of those terms that belong to the R-posterity of some term to which x has the relation R, in the sense which we gave before to "posterity," which includes a term in its own posterity. Reverting to the fundamental definitions, we find that the "proper posterity" may be defined as follows:—

The "proper posterity" of x with respect to R consists of all terms that possess every R-hereditary property possessed by every term to which x has the relation R.

It is to be observed that this definition has to be so framed as to be applicable not only when there is only one term to which x has the relation R, but also in cases (as *e.g.* that of father and child) where there may be many terms to which x has the relation R. We define further:

A term x is a "proper ancestor" of y with respect to R if y belongs to the proper posterity of x with respect to R.

We shall speak for short of "R-posterity" and "R-ancestors" when these terms seem more convenient.

Reverting now to the generation of series by the relation R between consecutive terms, we see that, if this method is to be possible, the relation "proper R-ancestor" must be an aliorelative, transitive, and connected. Under what circumstances will this occur? It will always be transitive: no matter what sort of relation R may be, "R-ancestor" and "proper R-ancestor" are always both transitive. But it is only under certain circumstances that it will be an aliorelative or connected. Consider, for example, the relation to one's left-hand neighbour at a round dinner-table at which there are twelve people. If we call this relation R, the proper R-posterity of a person consists of all who can be reached by going round the table from right to left. This includes everybody at the table, including the person himself, since twelve steps bring us back to our starting-point. Thus in such a case, though the relation "proper R-ancestor" is connected, and though R itself is an aliorelative, we do not get a series because "proper R-ancestor" is not an aliorelative. It is for this reason that we cannot say that one person comes before another with respect to the relation "right of" or to its ancestral derivative.

The above was an instance in which the ancestral relation was connected but not contained in diversity. An instance where it is contained in diversity but not connected is derived from the ordinary sense of the word "ancestor." If x is a proper ancestor of y, x and y cannot be the same person; but it is not true that of any two persons one must be an ancestor of the other.

The question of the circumstances under which series can be generated by ancestral relations derived from relations of consecutiveness is often important. Some of the most important cases are the following: Let R be a many-one relation, and let us confine our attention to the posterity of some term x. When so confined, the relation "proper R-ancestor" must be connected; therefore all that remains to ensure its being serial is that it shall be contained in diversity. This is a generalization of the instance of the dinner-table. Another generalization consists in taking R to be a one-one relation, and including the ancestry of x as well as the posterity. Here again, the one condition required to secure the generation of a series is that the relation "proper R-ancestor" shall be contained in diversity.

The generation of order by means of relations of consecutiveness, though important in its own sphere, is less general than the method which uses a transitive relation to define the order. It often happens in a series that there are an infinite number of intermediate terms between any two that may be selected, however near together these may be. Take, for instance, fractions in order of magnitude. Between any two fractions there are others—for example, the arithmetic mean of the two. Consequently there is no such thing as a pair of consecutive fractions. If we depended upon consecutiveness for defining order, we should not be able to define the order of magnitude among fractions. But in fact the relations of greater and less among fractions do not demand generation from relations of consecutiveness, and the relations of greater and less among fractions have the three characteristics which we need for defining serial relations. In all such cases the order must be defined by means of a *transitive* relation, since only such a relation is able to leap over an infinite number of intermediate terms. The method of consecutiveness, like that of counting for discovering the number of a collection, is appropriate to the finite; it may even be extended to certain infinite series, namely, those in which, though the total number of terms is infinite, the number of terms between any two is always finite; but it must not be regarded as general. Not only so, but care must be taken to eradicate from the imagination all habits of thought resulting from supposing it general. If this is not done, series in which there are no consecutive terms will remain difficult and puzzling. And such series are of vital importance for the understanding of continuity, space, time, and motion.

There are many ways in which series may be generated, but all depend upon the finding or construction of an asymmetrical transitive connected relation. Some of these ways have considerable importance. We may take as illustrative the generation of series by means of a three-term relation which we may call "between." This method is very useful in geometry, and may serve as an introduction to relations having more than two terms; it is best introduced in connection with elementary geometry.

Given any three points on a straight line in ordinary space, there must be one of them which is *between* the other two. This will not be the case with the points on a circle or any other closed curve, because, given any three points on a circle, we can travel from any one to any other without passing through the third. In fact, the notion "between" is characteristic of open series—or series in the strict sense—as opposed to what may be called "cyclic" series, where, as with people at the dinner-table, a sufficient journey brings us back to our starting-point. This notion of "between" may be chosen as the fundamental notion of ordinary geometry; but for the present we will only consider its application to a single straight line and to the ordering of the points on a straight line. [11] Taking any two points *a, b*, the line (*ab*) consists of three parts (besides *a* and *b* themselves):

[11] Cf. *Rivista di Matematica*, iv. pp. 55ff.; *Principles of Mathematics*, *p*. 394 (§375).

(1) Points between *a* and *b*.
(2) Points *x* such that *a* is between *x* and *b*.
(3) Points *y* such that *b* is between *y* and *a*.

Thus the line (*ab*) can be defined in terms of the relation "between."

In order that this relation "between" may arrange the points of the line in an order from left to right, we need certain assumptions, namely, the following:—

(1) If anything is between *a* and *b*, a and *b* are not identical.

(2) Anything between *a* and *b* is also between *b* and a.

(3) Anything between *a* and *b* is not identical with *a* (nor, consequently, with *b*, in virtue of (2)).

(4) If *x* is between *a* and *b*, anything between a and *x* is also between a and *b*.

(5) If *x* is between *a* and *b*, and *b* is between *x* and *y*, then *b* is between a and *y*.

(6) If *x* and *y* are between *a* and *b*, then either *x* and *y* are identical, or *x* is between *a*

and *y*, or *x* is between *y* and *b*.

(7) If *b* is between *a* and *x* and also between a and *y*, then either *x* and *y* are identical, or *x* is between *b* and *y*, or *y* is between *b* and *x*.

These seven properties are obviously verified in the case of points on a straight line in ordinary space. Any three-term relation which verifies them gives rise to series, as may be seen from the following definitions. For the sake of definiteness, let us assume that a is to the left of *b*. Then the points of the line (*ab*) are (1) those between which and *b*, a lies—these we will call to the left of *a*; (2) a itself; (3) those between a and *b*; (4) *b* itself; (5) those between which and a lies *b*—these we will call to the right of *b*. We may now define generally that of two points *x*, *y*, on the line (*ab*), we shall say that *x* is "to the left of" *y* in any of the following cases:—

(1) When *x* and *y* are both to the left of *a*, and *y* is between *x* and *a*;
(2) When *x* is to the left of *a*, and *y* is *a* or *b* or between a and *b* or to the right of *b*;
(3) When *x* is a, and *y* is between *a* and *b* or is *b* or is to the right of *b*;
(4) When *x* and *y* are both between *a* and *b*, and *y* is between *x* and *b*;
(5) When *x* is between *a* and *b*, and *y* is *b* or to the right of *b*;
(6) When *x* is *b* and *y* is to the right of *b*;
(7) When *x* and *y* are both to the right of *b* and *x* is between *b* and *y*.

It will be found that, from the seven properties which we have assigned to the relation "between," it can be deduced that the relation "to the left of," as above defined, is a *serial* relation as we defined that term. It is important to notice that nothing in the definitions or the argument depends upon our meaning by "between" the actual relation of that name which occurs in empirical space: any three-term relation having the above seven purely formal properties will serve the purpose of the argument equally well.

Cyclic order, such as that of the points on a circle, cannot be generated by means of three-term relations of "between." We need a relation of four terms, which may be called "separation of couples." The point may be illustrated by considering a journey round the world. One may go from England to New Zealand by way of Suez or by way of San Francisco; we cannot say definitely that either of these two places is "between" England and New Zealand. But if a man chooses that route to go round the world, whichever way round he goes, his times in England and New Zealand are separated from each other by his times in Suez and San Francisco, and conversely. Generalizing, if we take any four points on a circle, we can separate them into two couples, say *a* and *b* and *x* and *y*, such that, in order to get from *a* to *b* one must pass through either *x* or *y*, and in order to get from *x* to *y* one must pass through either a or *b*. Under these circumstances we say that the couple (*a*, *b*) are "separated" by the couple (*x*, *y*). Out of this relation a cyclic order can be generated, in a way resembling that in which we generated an open order from "between," but somewhat more complicated. [12]

[12] Cf. *Principles of Mathematics*, *p.* 205 (§194), and references there given.

The purpose of the latter half of this chapter has been to suggest the subject which one may call "generation of serial relations." When such relations have been defined, the generation of them from other relations possessing only some of the properties required for series becomes very important, especially in the philosophy of geometry and physics. But we cannot, within the limits of the present volume, do more than make the reader aware that such a subject exists.

CHAPTER V

KINDS OF RELATIONS

A great part of the philosophy of mathematics is concerned with *relations*, and many different kinds of relations have different kinds of uses. It often happens that a property which belongs to *all* relations is only important as regards relations of certain sorts; in these cases the reader will not see the bearing of the proposition asserting such a property unless he has in mind the sorts of relations for which it is useful. For reasons of this description, as well as from the intrinsic interest of the subject, it is well to have in our minds a rough list of the more mathematically serviceable varieties of relations.

We dealt in the preceding chapter with a supremely important class, namely, *serial* relations. Each of the three properties which we combined in defining series—namely, *asymmetry, transitiveness*, and *connexity*—has its own importance. We will begin by saying something on each of these three.

Asymmetry, *i.e.* the property of being incompatible with the converse, is a characteristic of the very greatest interest and importance. In order to develop its functions, we will consider various examples. The relation husband is asymmetrical, and so is the relation wife; *i.e.* if *a* is *husband* of *b*, *b* cannot be husband of *a*, and similarly in the case of *wife*. On the other hand, the relation "spouse" is symmetrical: if *a* is spouse of *b*, then *b* is spouse of *a*. Suppose now we are given the relation *spouse*, and we wish to derive the relation *husband*. *Husband* is the same as *male spouse* or *spouse of a female*; thus the relation *husband* can be derived from *spouse* either by limiting the domain to males or by limiting the converse domain to females. We see from this instance that, when a symmetrical relation is given, it is sometimes possible, without the help of any further relation, to separate it into two asymmetrical relations. But the cases where this is possible are rare and exceptional: they are cases where there are two mutually exclusive classes, say α and β, such that whenever the relation holds between two terms, one of the terms is a member of α and the other is a member of β—as, in the case of *spouse*, one term of the relation belongs to the class of males and one to the class of females. In such a case, the relation with its domain confined to α will be asymmetrical, and so will the relation with its domain confined to β. But such cases are not of the sort that occur when we are dealing with series of more than two terms; for in a series, all terms, except the first and last (if these exist), belong both to the domain and to the converse domain of the generating relation, so that a relation like *husband*, where the domain and converse domain do not overlap, is excluded.

The question how to *construct* relations having some useful property by means of operations upon relations which only have rudiments of the property is one of considerable importance. Transitiveness and connexity are easily constructed in many cases where the originally given relation does not possess them: for example, if R is any relation whatever, the ancestral relation derived from R by generalized induction is transitive; and if R is a many-one relation, the ancestral relation will be connected if confined to the posterity of a given term. But asymmetry is a much more difficult property to secure by construction. The method by which we derived *husband* from *spouse* is, as we have seen, not available in the most important cases, such as *greater*, *before, to the right of*, where domain and converse domain overlap. In all these cases, we can of course obtain a symmetrical relation by adding together the given relation and its converse, but we cannot pass back from this symmetrical relation to the original asymmetrical relation except by the help of some asymmetrical relation. Take, for example, the relation *greater*: the relation *greater or less—i.e. unequal*—is symmetrical, but there is nothing in this relation to show that it is the sum of two asymmetrical relations. Take such a relation as "differing in shape." This is not the sum of an asymmetrical relation and its converse, since shapes do not form a single series; but there is nothing to show that it differs from "differing in magnitude" if we did not already know that magnitudes have relations of greater and less. This illustrates the fundamental character of asymmetry as a property of relations.

From the point of view of the classification of relations, being asymmetrical is a much more important characteristic than implying diversity. Asymmetrical relations imply diversity, but the converse is not the case. "Unequal," for example, implies diversity, but is symmetrical. Broadly speaking, we may say that, if we wished as far as possible to dispense with relational propositions and replace them by such as ascribed predicates to subjects, we could succeed in this so long as we confined ourselves to *symmetrical* relations: those that do not imply diversity, if they are transitive, may be regarded as asserting a common predicate, while those that do imply diversity may be regarded as asserting incompatible predicates. For example, consider the relation of *similarity between classes*, by means of which we defined numbers. This relation is symmetrical and transitive and does not imply diversity. It would be possible, though less simple than the procedure we adopted, to regard the number of a collection as a predicate of the collection: then two similar classes will be two that have the same numerical predicate, while two that are not similar will be two that have different numerical predicates. Such a method of replacing relations by predicates is formally possible (though often very inconvenient) so long as the relations concerned are symmetrical; but it is formally impossible when the relations are asymmetrical, because both sameness and difference of predicates are symmetrical. Asymmetrical relations are, we may say, the most characteristically relational of relations, and the most important to the philosopher who wishes to study the ultimate logical nature of relations.

Another class of relations that is of the greatest use is the class of one-many relations, *i.e.* relations which at most one term can have to a given term. Such are father, mother, husband (except in Tibet), square of, sine of, and so on. But parent, square root, and so on, are not one-many. It is possible, formally, to replace all relations by one-many relations by means of a device. Take (say) the relation *less* among the inductive numbers. Given any number n greater than 1, there will not be only one number having the relation less to n, but we can form the whole class of numbers that are less than n. This is one class, and its relation to n is not shared by any other class. We may call the class of numbers that are less than n the "proper ancestry" of n, in the sense in which we spoke of ancestry and posterity in connection with mathematical induction. Then "proper ancestry" is a one-many relation (*one-many* will always be used so as to include *one-one*), since each number determines a single class of numbers as constituting its proper ancestry. Thus the relation *less than* can be replaced by *being a member of the proper ancestry of.* In this way a one-many relation in which the one is a class, together with membership of this class, can always formally replace a relation which is not one-many. Peano, who for some reason always instinctively conceives of a relation as one-many, deals in this way with those that are naturally not so. Reduction to one-many relations by this method, however, though possible as a matter of form, does not represent a technical simplification, and there is every reason to think that it does not represent a philosophical analysis, if only because classes must be regarded as "logical fictions." We shall therefore continue to regard one-many relations as a special kind of relations.

One-many relations are involved in all phrases of the form "the so-and-so of such-and-such." "The King of England," "the wife of Socrates," "the father of John Stuart Mill," and so on, all describe some person by means of a one-many relation to a given term. A person cannot have more than one father, therefore "the father of John Stuart Mill" described some one person, even if we did not know whom. There is much to say on the subject of descriptions, but for the present it is relations that we are concerned with, and descriptions are only relevant as exemplifying the uses of one-many relations. It should be observed that all mathematical functions result from one-many relations: the logarithm of x, the cosine of x, etc., are, like the father of x, terms described by means of a one-many relation (logarithm, cosine, etc.) to a given term (x). The notion of *function* need not be confined to numbers, or to the uses to which mathematicians have accustomed us; it can be extended to all cases of one-many relations, and "the father of x" is just as legitimately a function of which x is the argument as is "the logarithm of x." Functions in this sense are *descriptive* functions. As we shall see later, there are functions of a still more general and more fundamental sort, namely, *propositional* functions; but for the present we shall confine our attention to descriptive functions, *i.e.* "the term having the relation R to x," or, for short, "the R of x," where R is any one-many relation.

It will be observed that if "the R of x" is to describe a definite term, x must be a term to which something has the relation R, and there must not be more than one term having the relation R to x, since "the," correctly used, must imply uniqueness. Thus we may speak of "the father of x" if x is any human being except Adam and Eve; but we cannot speak of "the father of x" if x is a table or a chair or anything else that does not have a father. We shall say that the R of x "exists" when there is just one term, and no more, having the relation R to x. Thus if R is a one-many relation, the R of x exists whenever x belongs to the converse domain of R, and not otherwise. Regarding "the R of x" as a function in the mathematical sense, we say that x is the "argument" of the function, and if y is the term which has the relation R to x, *i.e.* if y is the R of x, then y is the "value" of the function for the argument x. If R is a one-many relation, the range of possible arguments to the function is the converse domain of R, and the range of values is the domain. Thus the range of possible arguments to the function "the father of x" is all who have fathers, *i.e.* the converse domain of the relation *father*, while the range of possible values for the function is all fathers, *i.e.* the domain of the relation.

Many of the most important notions in the logic of relations are descriptive functions, for example: *converse, domain, converse domain, field.* Other examples will occur as we proceed.

Among one-many relations, *one-one* relations are a specially important class. We have already had occasion to speak of one-one relations in connection with the definition of number, but it is necessary to be familiar with them, and not merely to know their formal definition. Their formal definition may be derived from that of one-many relations: they may be defined as one-many relations which are also the converses of one-many relations, *i.e.* as relations which are both one-many and many-one. One-many relations may be defined as relations such that, if x has the relation in question to y, there is no other term x' which also has the relation to y. Or, again, they may be defined as follows: Given two terms x and x', the terms to which x has the given relation and those to which x' has it have no member in common. Or, again, they may be defined as relations such that the relative product of one of them and its converse implies identity, where the "relative product" of two relations R and S is that relation which holds between *x* and *z* when there is an intermediate term *y*, such that *x* has the relation R to *y* and *y* has the relation S to *z*. Thus, for example, if R is the relation of father to son, the relative product of R and its converse will be the relation which holds between *x* and a man *z* when there is a person *y*, such that *x* is the father of *y* and *y* is the son of *z*. It is obvious that *x* and *z* must be the same person. If, on the other hand, we take the relation of parent and child, which is not one-many, we can no longer argue that, if *x* is a parent of *y* and *y* is a child of *z*, *x* and *z* must be the same person, because one may be the father of *y* and the other the mother. This illustrates that it is characteristic of one-many relations when the relative product of a relation and its converse implies identity. In the case of one-one relations this happens, and also the relative product of the converse and the relation implies identity. Given a relation R, it is convenient, if *x* has the relation R to *y*, to think of *y* as being reached from *x* by an "R-step" or an "R-vector." In the same case *x* will be reached from *y* by a "backward R-step." Thus we may state the characteristic of one-many relations with which we have been dealing by saying that an R-step followed by a backward R-step must bring us back to our starting-point. With other relations, this is by no means the case; for example, if R is the relation of child to parent, the relative product of R and its converse is the relation "self or brother or sister," and if R is the relation of grandchild to grandparent, the relative product of R and its converse is "self or brother or sister or first cousin." It will be observed that the relative product of two relations is not in general commutative, *i.e.* the relative product of R and S is not in general the same relation as the relative product of S and R. *E.g.* the relative product of parent and brother is uncle, but the relative product of brother and parent is parent.

One-one relations give a correlation of two classes, term for term, so that each term in either class has its correlate in the other. Such correlations are simplest to grasp when the two classes have no members in common, like the class of husbands and the class of wives; for in that case we know at once whether a term is to be considered as one *from* which the correlating relation R goes, or as one to which it goes. It is convenient to use the word *referent* for the term *from* which the relation goes, and the term *relatum* for the term *to* which it goes. Thus if *x* and *y* are husband and wife, then, with respect to the relation "husband," *x* is referent and *y* relatum, but with respect to the relation "wife," *y* is referent and *x* relatum. We say that a relation and its converse have opposite "senses"; thus the "sense" of a relation that goes from *x* to *y* is the opposite of that of the corresponding relation from *y* to *x*. The fact that a relation has a "sense" is fundamental, and is part of the reason why order can be generated by suitable relations. It will be observed that the class of all possible referents to a given relation is its domain, and the class of all possible relata is its converse domain.

But it very often happens that the domain and converse domain of a one-one relation overlap. Take, for example, the first ten integers (excluding 0), and add 1 to each; thus instead of the first ten integers we now have the integers

2, 3, 4, 5, 6, 7, 8, 9, 10, 11.

These are the same as those we had before, except that 1 has been cut off at the beginning and 11 has been joined on at the end. There are still ten integers: they are correlated with the previous ten by the relation of *n* to *n*+1, which is a one-one relation. Or, again, instead of adding 1 to each of our original ten integers, we could have doubled each of them, thus obtaining the integers

2, 4, 6, 8, 10, 12, 14, 16, 18, 20.

Here we still have five of our previous set of integers, namely, 2, 4, 6, 8, 10. The correlating relation in this case is the relation of a number to its double, which is again a one-one relation. Or we might have replaced each number by its square, thus obtaining the set

1, 4, 9, 16, 25, 36, 49, 64, 81, 100.

On this occasion only three of our original set are left, namely, 1, 4, 9. Such processes of correlation may be varied endlessly.

The most interesting case of the above kind is the case where our one-one relation has a converse domain which is part, but not the whole, of the domain. If, instead of confining the domain to the first ten integers, we had considered the whole of the inductive numbers, the above instances would have illustrated this case. We may place the numbers concerned in two rows, putting the correlate directly under the number whose correlate it is. Thus when the correlator is the relation of n to $n+1$, we have the two rows:

1, 2, 3, 4, 5, ... n ...
2, 3, 4, 5, 6, ... $n+1$ ...

When the correlator is the relation of a number to its double, we have the two rows:

1, 2, 3, 4, 5, ... n ...
2, 4, 6, 8, 10, ... 2n ...

When the correlator is the relation of a number to its square, the rows are:

1, 2, 3, 4, 5, ... n ...
1, 4, 9, 16, 25, ... n2 ...

In all these cases, all inductive numbers occur in the top row, and only some in the bottom row.

Cases of this sort, where the converse domain is a "proper part" of the domain (*i.e.* a part not the whole), will occupy us again when we come to deal with infinity. For the present, we wish only to note that they exist and demand consideration.

Another class of correlations which are often important is the class called "permutations," where the domain and converse domain are identical. Consider, for example, the six possible arrangements of three letters:

a, b, c
a, c, b
b, c, a
b, a, c
c, a, b
c, b, a

Each of these can be obtained from any one of the others by means of a correlation. Take, for example, the first and last, (a, b, *c*) and (*c*, b, a). Here *a* is correlated with *c*, *b* with itself, and *c* with a. It is obvious that the combination of two permutations is again a permutation, *i.e.* the permutations of a given class form what is called a "group."

These various kinds of correlations have importance in various connections, some for one purpose, some for another. The general notion of one-one correlations has boundless importance in the philosophy of mathematics, as we have partly seen already, but shall see much more fully as we proceed. One of its uses will occupy us in our next chapter.

CHAPTER VI

SIMILARITY OF RELATIONS

We saw in Chapter II. that two classes have the same number of terms when they are "similar," *i.e.* when there is a one-one relation whose domain is the one class and whose converse domain is the other. In such a case we say that there is a "one-one correlation" between the two classes.

In the present chapter we have to define a relation between relations, which will play the same part for them that similarity of classes plays for classes. We will call this relation "similarity of relations," or "likeness" when it seems desirable to use a different word from that which we use for classes. How is likeness to be defined?

We shall employ still the notion of correlation: we shall assume that the domain of the one relation can be correlated with the domain of the other, and the converse domain with the converse domain; but that is not enough for the sort of resemblance which we desire to have between our two relations. What we desire is that, whenever either relation holds between two terms, the other relation shall hold between the correlates of these two terms. The easiest example of the sort of thing we desire is a map. When one place is north of another, the place on the map corresponding to the one is above the place on the map corresponding to the other; when one place is west of another, the place on the map corresponding to the one is to the left of the place on the map corresponding to the other; and so on. The structure of the map corresponds with that of the country of which it is a map. The space-relations in the map have "likeness" to the space-relations in the country mapped. It is this kind of connection between relations that we wish to define.

We may, in the first place, profitably introduce a certain restriction. We will confine ourselves, in defining likeness, to such relations as have "fields," *i.e.* to such as permit of the formation of a single class out of the domain and the converse domain. This is not always the case. Take, for example, the relation "domain," *i.e.* the relation which the domain of a relation has to the relation. This relation has all classes for its domain, since every class is the domain of some relation; and it has all relations for its converse domain, since every relation has a domain. But classes and relations cannot be added together to form a new single class, because they are of different logical "types." We do not need to enter upon the difficult doctrine of types, but it is well to know when we are abstaining from entering upon it. We may say, without entering upon the grounds for the assertion, that a relation only has a "field" when it is what we call "homogeneous," *i.e.* when its domain and converse domain are of the same logical type; and as a rough-and-ready indication of what we mean by a "type," we may say that individuals, classes of individuals, relations between individuals, relations between classes, relations of classes to individuals, and so on, are different types. Now the notion of likeness is not very useful as applied to relations that are not homogeneous; we shall, therefore, in defining likeness, simplify our problem by speaking of the "field" of one of the relations concerned. This somewhat limits the generality of our definition, but the limitation is not of any practical importance. And having been stated, it need no longer be remembered. We may define two relations P and Q as "similar," or as having "likeness," when there is a one-one relation S whose domain is the field of P and whose converse domain is the field of Q, and which is such that, if one term has the relation P to another, the correlate of the one has the relation Q to the correlate of the other, and vice versa. A figure will make this clearer.

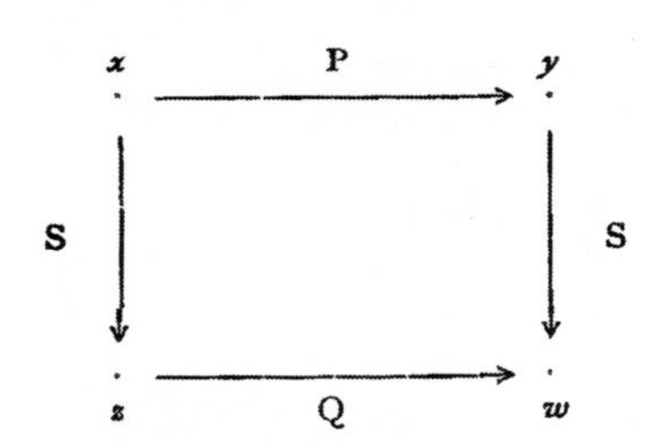

Let x and y be two terms having the relation P. Then there are to be two terms z, w, such that x has the relation S to z, y has the relation S to w, and z has the relation Q to w. If this happens with every pair of terms such as x and y, and if the converse happens with every pair of terms such as z and w, it is clear that for every instance in which the relation P holds there is a corresponding instance in which the relation Q holds, and *vice versa*; and this is what we desire to secure by our definition. We can eliminate some redundancies in the above sketch of a definition, by observing that, when the above conditions are realized, the relation P is the same as the relative product of S and Q and the converse of S, *i.e.* the P-step from x to y may be replaced by the succession of the S-step from x to z, the Q-step from z to w, and the backward S-step from w to y. Thus we may set up the following definitions:—

A relation S is said to be a "correlator" or an "ordinal correlator" of two relations P and Q if S is one-one, has the field of Q for its converse domain, and is such that P is the relative product of S and Q and the converse of S.

Two relations P and Q are said to be "similar," or to have "likeness," when there is at least one correlator of P and Q.

These definitions will be found to yield what we above decided to be necessary.

It will be found that, when two relations are similar, they share all properties which do not depend upon the actual terms in their fields. For instance, if one implies diversity, so does the other; if one is transitive, so is the other; if one is connected, so is the other.

Hence if one is serial, so is the other. Again, if one is one-many or one-one, the other is one-many or one-one; and so on, through all the general properties of relations. Even statements involving the actual terms of the field of a relation, though they may not be true as they stand when applied to a similar relation, will always be capable of translation into statements that are analogous. We are led by such considerations to a problem which has, in mathematical philosophy, an importance by no means adequately recognised hitherto. Our problem may be stated as follows:—

Given some statement in a language of which we know the grammar and the syntax, but not the vocabulary, what are the possible meanings of such a statement, and what are the meanings of the unknown words that would make it true?

The reason that this question is important is that it represents, much more nearly than might be supposed, the state of our knowledge of nature. We know that certain scientific propositions—which, in the most advanced sciences, are expressed in mathematical symbols—are more or less true of the world, but we are very much at sea as to the interpretation to be put upon the terms which occur in these propositions. We know much more (to use, for a moment, an old-fashioned pair of terms) about the *form* of nature than about the *matter*. Accordingly, what we really know when we enunciate a law of nature is only that there is probably *some* interpretation of our terms which will make the law approximately true. Thus great importance attaches to the question: What are the possible meanings of a law expressed in terms of which we do not know the substantive meaning, but only the grammar and syntax? And this question is the one suggested above.

For the present we will ignore the general question, which will occupy us again at a later stage; the subject of likeness itself must first be further investigated.

Owing to the fact that, when two relations are similar, their properties are the same except when they depend upon the fields being composed of just the terms of which they are composed, it is desirable to have a nomenclature which collects together all the relations that are similar to a given relation. Just as we called the set of those classes that are similar to a given class the "number" of that class, so we may call the set of all those relations that are similar to a given relation the "number" of that relation. But in order to avoid confusion with the numbers appropriate to classes, we will speak, in this case, of a "relation-number." Thus we have the following definitions:—

The "relation-number" of a given relation is the class of all those relations that are similar to the given relation.

"Relation-numbers" are the set of all those classes of relations that are relation-numbers of various relations; or, what comes to the same thing, a relation-number is a class of relations consisting of all those relations that are similar to one member of the class.

When it is necessary to speak of the numbers of classes in a way which makes it impossible to confuse them with relation-numbers, we shall call them "cardinal numbers." Thus cardinal numbers are the numbers appropriate to classes. These include the ordinary integers of daily life, and also certain infinite numbers, of which we shall speak later. When we speak of "numbers" without qualification, we are to be understood as meaning *cardinal* numbers. The definition of a cardinal number, it will be remembered, is as follows:—

The "cardinal number" of a given class is the set of all those classes that are similar to the given class.

The most obvious application of relation-numbers is to *series*. Two series may be regarded as equally long when they have the same relation-number. Two *finite* series will have the same relation-number when their fields have the same cardinal number of terms, and only then—*i.e.* a series of (say) 15 terms will have the same relation-number as any other series of fifteen terms, but will not have the same relation-number as a series of 14 or 16 terms, nor, of course, the same relation-number as a relation which is not serial. Thus, in the quite special case of finite series, there is parallelism between cardinal and relation-numbers. The relation-numbers applicable to series may be called "serial numbers" (what are commonly called "ordinal numbers" are a sub-class of these); thus a finite serial number is determinate when we know the cardinal number of terms in the field of a series having the serial number in question. If n is a finite cardinal number, the relation-number of a series which has n terms is called the "ordinal" number n. (There are also infinite ordinal numbers, but of them we shall speak in a later chapter.) When the cardinal number of terms in the field of a series is infinite, the relation-number of the series is not determined merely by the cardinal number, indeed an infinite number of relation-numbers exist for one infinite cardinal number, as we shall see when we come to consider infinite series. When a series is infinite, what we may call its "length," *i.e.* its relation-number, may vary without change in the cardinal number; but when a series is finite, this cannot happen.

We can define addition and multiplication for relation-numbers as well as for cardinal numbers, and a whole arithmetic of relation-numbers can be developed. The manner in which this is to be done is easily seen by considering the case of series. Suppose, for example, that we wish to define the sum of two non-overlapping series in such a way that the relation-number of the sum shall be capable of being defined as the sum of the relation-numbers of the two series. In the first place, it is clear that there is an *order* involved as between the two series: one of them must be placed before the other. Thus if P and Q are the generating relations of the two series, in the series which is their sum with P put before Q, every member of the field of P will precede every member of the field of Q. Thus the serial relation which is to be defined as the sum of P and Q is not "P or Q" simply, but "P or Q or the relation of any member of the field of P to any member of the field of Q." Assuming that P and Q do not overlap, this relation is serial, but "P or Q" is not serial, being not connected, since it does not hold between a member of the field of P and a member of the field of Q. Thus the sum of P and Q, as above defined, is what we need in order to define the sum of two relation-numbers. Similar modifications are needed for products and powers. The resulting arithmetic does not obey the commutative law: the sum or product of two relation-numbers generally depends upon the order in which they are taken. But it obeys the associative law, one form of the distributive law, and two of the formal laws for powers, not only as applied to serial numbers, but as applied to relation-numbers generally. Relation-arithmetic, in fact, though recent, is a thoroughly respectable branch of mathematics.

It must not be supposed, merely because series afford the most obvious application of the idea of likeness, that there are no other applications that are important. We have already mentioned maps, and we might extend our thoughts from this illustration to geometry generally. If the system of relations by which a geometry is applied to a certain set of terms can be brought fully into relations of likeness with a system applying to another set of terms, then the geometry of the two sets is indistinguishable from the mathematical point of view, *i.e.* all the propositions are the same, except for the fact that they are applied in one case to one set of terms and in the other to another. We may illustrate this by the relations of the sort that may be called "between," which we considered in Chapter IV. We there saw that, provided a three-term relation has certain formal logical properties, it will give rise to series, and may be called a "between-relation." Given any two points, we can use the between-relation to define the straight line determined by those two points; it consists of a and *b* together with all points *x*, such that the between-relation holds between the three points a, *b*, *x* in some order or other. It has been shown by O. Veblen that we may regard our whole space as the field of a three-term between-relation, and define our geometry by the properties we assign to our between-relation. [13] Now likeness is just as easily definable between three-term relations as between two-term relations. If B and B' are two between-relations, so that "*x*B(*y*, *z*)" means "*x* is between *y* and *z* with respect to B," we shall call S a correlator of B and B' if it has the field of B' for its converse domain, and is such that the relation B holds between three terms when B' holds between their S-correlates, and only then. And we shall say that B is like B' when there is at least one correlator of B with B'. The reader can easily convince himself that, if B is like B' in this sense, there can be no difference between the geometry generated by B and that generated by B'.

[13] This does not apply to elliptic space, but only to spaces in which the straight line is an open series. *Modern Mathematics*, edited by J. W. A. Young, pp. 3–51 (monograph by O. Veblen on "The Foundations of Geometry").

It follows from this that the mathematician need not concern himself with the particular being or intrinsic nature of his points, lines, and planes, even when he is speculating as an *applied* mathematician. We may say that there is empirical evidence of the approximate truth of such parts of geometry as are not matters of definition. But there is no empirical evidence as to what a "point" is to be. It has to be something that as nearly as possible satisfies our axioms, but it does not have to be "very small" or "without parts." Whether or not it is those things is a matter of indifference, so long as it satisfies the axioms. If we can, out of empirical material, construct a logical structure, no matter how complicated, which will satisfy our geometrical axioms, that structure may legitimately be called a "point." We must not say that there is nothing else that could legitimately be called a "point"; we must only say: "This object we have constructed is sufficient for the geometer; it may be one of many objects, any of which would be sufficient, but that is no concern of ours, since this object is enough to vindicate the empirical truth of geometry, in so far as geometry is not a matter of definition." This is only an illustration of the general principle that what matters in mathematics, and to a very great extent in physical science, is not the intrinsic nature of our terms, but the logical nature of their interrelations.

We may say, of two similar relations, that they have the same "structure." For mathematical purposes (though not for those of pure philosophy) the only thing of importance about a relation is the cases in which it holds, not its intrinsic nature. Just as a class may be defined by various different but co-extensive concepts—*e.g.* "man" and "featherless biped"—so two relations which are conceptually different may hold in the same set of instances. An "instance" in which a relation holds is to be conceived as a couple of terms, with an order, so that one of the terms comes first and the other second; the couple is to be, of course, such that its first term has the relation in question to its second. Take (say) the relation "father": we can define what we may call the "extension" of this relation as the class of all ordered couples (*x*, *y*) which are such that *x* is the father of *y*. From the mathematical point of view, the only thing of importance about the relation "father" is that it defines this set of ordered couples. Speaking generally, we say:

The "extension" of a relation is the class of those ordered couples (x, y) which are such that x has the relation in question to y.

We can now go a step further in the process of abstraction, and consider what we mean by "structure." Given any relation, we can, if it is a sufficiently simple one, construct a map of it. For the sake of definiteness, let us take a relation of which the extension is the following couples: *ab, ac, ad, bc, ce, dc, de*, where *a, b, c, d, e* are five terms, no matter what. We may make a "map" of this relation by taking five points on a plane and connecting them by arrows, as in the accompanying figure.

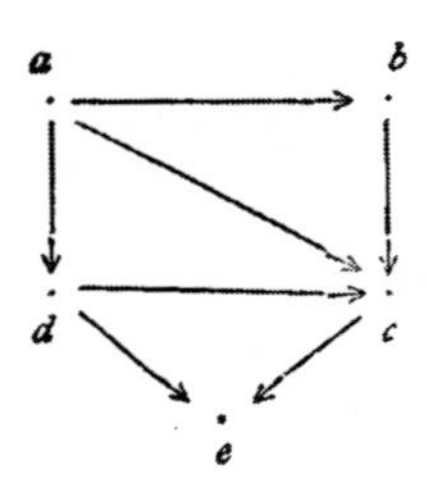

What is revealed by the map is what we call the "structure" of the relation.

It is clear that the "structure" of the relation does not depend upon the particular terms that make up the field of the relation. The field may be changed without changing the structure, and the structure may be changed without changing the field—for example, if we were to add the couple *ae* in the above illustration we should alter the structure but not the field. Two relations have the same "structure," we shall say, when the same map will do for both—or, what comes to the same thing, when either can be a map for the other (since every relation can be its own map). And that, as a moment's reflection shows, is the very same thing as what we have called "likeness." That is to say, two relations have the same structure when they have likeness, *i.e.* when they have the same relation-number. Thus what we defined as the "relation-number" is the very same thing as is obscurely intended by the word "structure"—a word which, important as it is, is never (so far as we know) defined in precise terms by those who use it.

There has been a great deal of speculation in traditional philosophy which might have been avoided if the importance of structure, and the difficulty of getting behind it, had been realized. For example, it is often said that space and time are subjective, but they have objective counterparts; or that phenomena are subjective, but are caused by things in themselves, which must have differences *inter se* corresponding with the differences in the phenomena to which they give rise. Where such hypotheses are made, it is generally supposed that we can know very little about the objective counterparts. In actual fact, however, if the hypotheses as stated were correct, the objective counterparts would form a world having the same structure as the phenomenal world, and allowing us to infer from phenomena the truth of all propositions that can be stated in abstract terms and are known to be true of phenomena. If the phenomenal world has three dimensions, so must the world behind phenomena; if the phenomenal world is Euclidean, so must the other be; and so on. In short, every proposition having a communicable significance must be true of both worlds or of neither: the only difference must lie in just that essence of individuality which always eludes words and baffles description, but which, for that very reason, is irrelevant to science. Now the only purpose that philosophers have in view in condemning phenomena is in order to persuade themselves and others that the real world is very different from the world of appearance. We can all sympathise with their wish to prove such a very desirable proposition, but we cannot congratulate them on their success. It is true that many of them do not assert objective counterparts to phenomena, and these escape from the above argument. Those who do assert counterparts are, as a rule, very reticent on the subject, probably because they feel instinctively that, if pursued, it will bring about too much of a *rapprochement* between the real and the phenomenal world. If they were to pursue the topic, they could hardly avoid the conclusions which we have been suggesting. In such ways, as well as in many others, the notion of structure or relation-number is important.

CHAPTER VII

RATIONAL, REAL, AND COMPLEX NUMBERS

We have now seen how to define cardinal numbers, and also relation-numbers, of which what are commonly called ordinal numbers are a particular species. It will be found that each of these kinds of number may be infinite just as well as finite. But neither is capable, as it stands, of the more familiar extensions of the idea of number, namely, the extensions to negative, fractional, irrational, and complex numbers. In the present chapter we shall briefly supply logical definitions of these various extensions.

One of the mistakes that have delayed the discovery of correct definitions in this region is the common idea that each extension of number included the previous sorts as special cases. It was thought that, in dealing with positive and negative integers, the positive integers might be identified with the original signless integers. Again it was thought that a fraction whose denominator is 1 may be identified with the natural number which is its numerator. And the irrational numbers, such as the square root of 2, were supposed to find their place among rational fractions, as being greater than some of them and less than the others, so that rational and irrational numbers could be taken together as one class, called "real numbers." And when the idea of number was further extended so as to include "complex" numbers, *i.e.* numbers involving the square root of -1, it was thought that real numbers could be regarded as those among complex numbers in which the imaginary part (*i.e.* the part which was a multiple of the square root of -1) was zero. All these suppositions were erroneous, and must be discarded, as we shall find, if correct definitions are to be given.

Let us begin with *positive and negative integers*. It is obvious on a moment's consideration that $+1$ and -1 must both be relations, and in fact must be each other's converses. The obvious and sufficient definition is that $+1$ is the relation of $n+1$ to n, and -1 is the relation of n to $n+1$. Generally, if m is any inductive number, $+m$ will be the relation of $n+m$ to n (for any n), and $-m$ will be the relation of n to $n+m$. According to this definition, $+m$ is a relation which is one-one so long as n is a cardinal number (finite or infinite) and m is an inductive cardinal number. But $+m$ is under no circumstances capable of being identified with m, which is not a relation, but a class of classes. Indeed, $+m$ is every bit as distinct from m as $-m$ is.

Fractions are more interesting than positive or negative integers. We need fractions for many purposes, but perhaps most obviously for purposes of measurement. My friend and collaborator Dr A. N. Whitehead has developed a theory of fractions specially adapted for their application to measurement, which is set forth in *Principia Mathematica*. [14] But if all that is needed is to define objects having the required purely mathematical properties, this purpose can be achieved by a simpler method, which we shall here adopt. We shall define the fraction m/n as being that relation which holds between two inductive numbers x, y when $xn=ym$. This definition enables us to prove that m/n is a one-one relation, provided neither m nor n is zero. And of course n/m is the converse relation to m/n.

[14] Vol. iii. *300ff., especially 303.

From the above definition it is clear that the fraction $m/1$ is that relation between two integers x and y which consists in the fact that $x=my$. This relation, like the relation $+m$, is by no means capable of being identified with the inductive cardinal number m, because a relation and a class of classes are objects of utterly different kinds. [15] It will be seen that $0/n$ is always the same relation, whatever inductive number n may be; it is, in short, the relation of 0 to any other inductive cardinal. We may call this the zero of rational numbers; it is not, of course, identical with the cardinal number 0. Conversely, the relation $m/0$ is always the same, whatever inductive number m may be. There is not any inductive cardinal to correspond to $m/0$. We may call it "the infinity of rationals." It is an instance of the sort of infinite that is traditional in mathematics, and that is represented by "∞." This is a totally different sort from the true Cantorian infinite, which we shall consider in our next chapter. The infinity of rationals does not demand, for its definition or use, any infinite classes or infinite integers. It is not, in actual fact, a very important notion, and we could dispense with it altogether if there were any object in doing so. The Cantorian infinite, on the other hand, is of the greatest and most fundamental importance; the understanding of it opens the way to whole new realms of mathematics and philosophy.

[15] Of course in practice we shall continue to speak of a fraction as (say) greater or less than 1, meaning greater or less than the ratio 1/1. So long as it is understood that the ratio 1/1 and the cardinal number 1 are different, it is not necessary to be always pedantic in emphasising the difference.

It will be observed that zero and infinity, alone among ratios, are not one-one. Zero is one-many, and infinity is many-one.

There is not any difficulty in defining *greater* and *less* among ratios (or fractions). Given two ratios m/n and p/q, we shall say that m/n is less than p/q if mq is less than pn. There is no difficulty in proving that the relation "less than," so defined, is serial, so that the ratios form a series in order of magnitude. In this series, zero is the smallest term and infinity is the largest. If we omit zero and infinity from our series, there is no longer any smallest or largest ratio; it is obvious that if m/n is any ratio other than zero and infinity, $m/2n$ is smaller and $2m/n$ is larger, though neither is zero or infinity, so that m/n is neither the smallest nor the largest ratio, and therefore (when zero and infinity are omitted) there is no smallest or largest, since m/n was chosen arbitrarily. In like manner we can prove that however nearly equal two fractions may be, there are always other fractions between them. For, let m/n and p/q be two fractions, of which p/q is the greater. Then it is easy to see (or to prove) that $(m+p)/(n+q)$ will be greater than m/n and less than p/q. Thus the series of ratios is one in which no two terms are consecutive, but there are always other terms between any two. Since there are other terms between these others, and so on *ad infinitum*, it is obvious that there are an infinite number of ratios between any two, however nearly equal these two may be. [16] A series having the property that there are always other terms between any two, so that no two are consecutive, is called "compact." Thus the ratios in order of magnitude form a "compact" series. Such series have many important properties, and it is important to observe that ratios afford an instance of a compact series generated purely logically, without any appeal to space or time or any other empirical datum.

[16] Strictly speaking, this statement, as well as those following to the end of the paragraph, involves what is called the "axiom of infinity," which will be discussed in a later chapter.

Positive and negative ratios can be defined in a way analogous to that in which we defined positive and negative integers. Having first defined the sum of two ratios m/n and p/q as $(mq+pn)/nq$, we define $+p/q$ as the relation of $m/n+p/q$ to m/n, where m/n is any ratio; and $-p/q$ is of course the converse of $+p/q$. This is not the only possible way of defining positive and negative ratios, but it is a way which, for our purpose, has the merit of being an obvious adaptation of the way we adopted in the case of integers.

We come now to a more interesting extension of the idea of number, *i.e.* the extension to what are called "real" numbers, which are the kind that embrace irrationals. In Chapter I. we had occasion to mention "incommensurables" and their discovery by Pythagoras. It was through them, *i.e.* through geometry, that irrational numbers were first thought of. A square of which the side is one inch long will have a diagonal of which the length is the square root of 2 inches. But, as the ancients discovered, there is no fraction of which the square is 2. This proposition is proved in the tenth book of Euclid, which is one of those books that schoolboys supposed to be fortunately lost in the days when Euclid was still used as a text-book. The proof is extraordinarily simple. If possible, let m/n be the square root of 2, so that $m^2/n^2 = 2$, *i.e.* $m^2 = 2n^2$. Thus m^2 is an even number, and therefore m must be an even number, because the square of an odd number is odd. Now if m is even, m^2 must divide by 4, for if $m=2p$, then $m^2=4p^2$. Thus we shall have $4p^2 = 2n^2$, where p is half of m. Hence $2p^2=n^2$, and therefore n/p will also be the square root of 2. But then we can repeat the argument: if $n=2q$, p/q will also be the square root of 2, and so on, through an unending series of numbers that are each half of its predecessor. But this is impossible; if we divide a number by 2, and then halve the half, and so on, we must reach an odd number after a finite number of steps. Or we may put the argument even more simply by assuming that the m/n we start with is in its lowest terms; in that case, m and n cannot both be even; yet we have seen that, if $m^2/n^2=2$, they must be. Thus there cannot be any fraction m/n whose square is 2.

Thus no fraction will express exactly the length of the diagonal of a square whose side is one inch long. This seems like a challenge thrown out by nature to arithmetic. However the arithmetician may boast (as Pythagoras did) about the power of numbers, nature seems able to baffle him by exhibiting lengths which no numbers can estimate in terms of the unit. But the problem did not remain in this geometrical form. As soon as algebra was invented, the same problem arose as regards the solution of equations, though here it took on a wider form, since it also involved complex numbers.

It is clear that fractions can be found which approach nearer and nearer to having their square equal to 2. We can form an ascending series of fractions all of which have their squares less than 2, but differing from 2 in their later members by less than any assigned amount. That is to say, suppose I assign some small amount in advance, say one-billionth, it will be found that all the terms of our series after a certain one, say the tenth, have squares that differ from 2 by less than this amount. And if I had assigned a still smaller amount, it might have been necessary to go further along the series, but we should have reached sooner or later a term in the series, say the twentieth, after which all terms would have had squares differing from 2 by less than this still smaller amount. If we set to work to extract the square root of 2 by the usual arithmetical rule, we shall obtain an unending decimal which, taken to so-and-so many places, exactly fulfils the above conditions. We can equally well form a descending series of fractions whose squares are all greater than 2, but greater by continually smaller amounts as we come to later terms of the series, and differing, sooner or later, by less than any assigned amount. In this way we seem to be drawing a cordon round the square root of 2, and it may seem difficult to believe that it can permanently escape us. Nevertheless, it is not by this method that we shall actually reach the square root of 2.

If we divide *all* ratios into two classes, according as their squares are less than 2 or not, we find that, among those whose squares are *not* less than 2, all have their squares greater than 2. There is no maximum to the ratios whose square is less than 2, and no minimum to those whose square is greater than 2. There is no lower limit short of zero to the difference between the numbers whose square is a little less than 2 and the numbers whose square is a little greater than 2. We can, in short, divide all ratios into two classes such that all the terms in one class are less than all in the other, there is no maximum to the one class, and there is no minimum to the other. Between these two classes, where $\sqrt{2}$ ought to be, there is nothing. Thus our cordon, though we have drawn it as tight as possible, has been drawn in the wrong place, and has not caught $\sqrt{2}$.

The above method of dividing all the terms of a series into two classes, of which the one wholly precedes the other, was brought into prominence by Dedekind, [17] and is therefore called a "Dedekind cut." With respect to what happens at the point of section, there are four possibilities: (1) there may be a maximum to the lower section and a minimum to the upper section, (2) there may be a maximum to the one and no minimum to the other, (3) there may be no maximum to the one, but a minimum to the other, (4) there may be neither a maximum to the one nor a minimum to the other. Of these four cases, the first is illustrated by any series in which there are consecutive terms: in the series of integers, for instance, a lower section must end with some number n and the upper section must then begin with $n+1$. The second case will be illustrated in the series of ratios if we take as our lower section all ratios up to and including 1, and in our upper section all ratios greater than 1. The third case is illustrated if we take for our lower section all ratios less than 1, and for our upper section all ratios from 1 upward (including 1 itself). The fourth case, as we have seen, is illustrated if we put in our lower section all ratios whose square is less than 2, and in our upper section all ratios whose square is greater than 2.

[17] *Stetigkeit und irrationale Zahlen*, 2nd edition, Brunswick, 1892.

We may neglect the first of our four cases, since it only arises in series where there are consecutive terms. In the second of our four cases, we say that the maximum of the lower section is the *lower limit* of the upper section, or of any set of terms chosen out of the upper section in such a way that no term of the upper section is before all of them. In the third of our four cases, we say that the minimum of the upper section is the upper limit of the lower section, or of any set of terms chosen out of the lower section in such a way that no term of the lower section is after all of them. In the fourth case, we say that there is a "gap": neither the upper section nor the lower has a limit or a last term. In this case, we may also say that we have an "irrational section," since sections of the series of ratios have "gaps" when they correspond to irrationals.

What delayed the true theory of irrationals was a mistaken belief that there must be "limits" of series of ratios. The notion of "limit" is of the utmost importance, and before proceeding further it will be well to define it.

A term x is said to be an "upper limit" of a class α with respect to a relation P if (1) α has no maximum in P, (2) every member of α which belongs to the field of P precedes x, (3) every member of the field of P which precedes x precedes some member of α. (By "precedes" we mean "has the relation P to.")

This presupposes the following definition of a "maximum":—

A term x is said to be a "maximum" of a class α with respect to a relation P if x is a member of α and of the field of P and does not have the relation P to any other member of α.

These definitions do not demand that the terms to which they are applied should be quantitative. For example, given a series of moments of time arranged by earlier and later, their "maximum" (if any) will be the last of the moments; but if they are arranged by later and earlier, their "maximum" (if any) will be the first of the moments.

The "minimum" of a class with respect to P is its maximum with respect to the converse of P; and the "lower limit" with respect to P is the upper limit with respect to the converse of P.

The notions of limit and maximum do not essentially demand that the relation in respect to which they are defined should be serial, but they have few important applications except to cases when the relation is serial or quasi-serial. A notion which is often important is the notion "upper limit or maximum," to which we may give the name "upper boundary." Thus the "upper boundary" of a set of terms chosen out of a series is their last member if they have one, but, if not, it is the first term after all of them, if there is such a term. If there is neither a maximum nor a limit, there is no upper boundary. The "lower boundary" is the lower limit or minimum.

Reverting to the four kinds of Dedekind section, we see that in the case of the first three kinds each section has a boundary (upper or lower as the case may be), while in the fourth kind neither has a boundary. It is also clear that, whenever the lower section has an upper boundary, the upper section has a lower boundary. In the second and third cases, the two boundaries are identical; in the first, they are consecutive terms of the series.

A series is called "Dedekindian" when every section has a boundary, upper or lower as the case may be.

We have seen that the series of ratios in order of magnitude is not Dedekindian.

From the habit of being influenced by spatial imagination, people have supposed that series *must* have limits in cases where it seems odd if they do not. Thus, perceiving that there was no *rational* limit to the ratios whose square is less than 2, they allowed themselves to "postulate" an *irrational* limit, which was to fill the Dedekind gap. Dedekind, in the above-mentioned work, set up the axiom that the gap must always be filled, *i.e.* that every section must have a boundary. It is for this reason that series where his axiom is verified are called "Dedekindian." But there are an infinite number of series for which it is not verified.

The method of "postulating" what we want has many advantages; they are the same as the advantages of theft over honest toil. Let us leave them to others and proceed with our honest toil.

It is clear that an irrational Dedekind cut in some way "represents" an irrational. In order to make use of this, which to begin with is no more than a vague feeling, we must find some way of eliciting from it a precise definition; and in order to do this, we must disabuse our minds of the notion that an irrational must be the limit of a set of ratios. Just as ratios whose denominator is 1 are not identical with integers, so those rational numbers which can be greater or less than irrationals, or can have irrationals as their limits, must not be identified with ratios. We have to define a new kind of numbers called "real numbers," of which some will be rational and some irrational. Those that are rational "correspond" to ratios, in the same kind of way in which the ratio $n/1$ corresponds to the integer n; but they are not the same as ratios. In order to decide what they are to be, let us observe that an irrational is represented by an irrational cut, and a cut is represented by its lower section. Let us confine ourselves to cuts in which the lower section has no maximum; in this case we will call the lower section a "segment." Then those segments that correspond to ratios are those that consist of all ratios less than the ratio they correspond to, which is their boundary; while those that represent irrationals are those that have no boundary. Segments, both those that have boundaries and those that do not, are such that, of any two pertaining to one series, one must be part of the other; hence they can all be arranged in a series by the relation of whole and part. A series in which there are Dedekind gaps, *i.e.* in which there are segments that have no boundary, will give rise to more segments than it has terms, since each term will define a segment having that term for boundary, and then the segments without boundaries will be extra.

We are now in a position to define a real number and an irrational number.

A "real number" is a segment of the series of ratios in order of magnitude.

An "irrational number" is a segment of the series of ratios which has no boundary.

A "rational real number" is a segment of the series of ratios which has a boundary.

Thus a rational real number consists of all ratios less than a certain ratio, and it is the rational real number corresponding to that ratio. The real number 1, for instance, is the class of proper fractions.

In the cases in which we naturally supposed that an irrational must be the limit of a set of ratios, the truth is that it is the limit of the corresponding set of rational real numbers in the series of segments ordered by whole and part. For example, $\sqrt{2}$ is the upper limit of all those segments of the series of ratios that correspond to ratios whose square is less than 2. More simply still, $\sqrt{2}$ is the segment *consisting* of all those ratios whose square is less than 2.

It is easy to prove that the series of segments of any series is Dedekindian. For, given any set of segments, their boundary will be their logical sum, *i.e.* the class of all those terms that belong to at least one segment of the set. [18]

[18] For a fuller treatment of the subject of segments and Dedekindian relations, see *Principia Mathematica*, vol. ii. *210–214. For a fuller treatment of real numbers, see *ibid.*, vol. iii. *310ff., and *Principles of Mathematics*, chaps. xxxiii. and xxxiv.

The above definition of real numbers is an example of "construction" as against "postulation," of which we had another example in the definition of cardinal numbers. The great advantage of this method is that it requires no new assumptions, but enables us to proceed deductively from the original apparatus of logic.

There is no difficulty in defining addition and multiplication for real numbers as above defined. Given two real numbers μ and ν, each being a class of ratios, take any member of μ and any member of ν and add them together according to the rule for the addition of ratios. Form the class of all such sums obtainable by varying the selected members of μ and ν. This gives a new class of ratios, and it is easy to prove that this new class is a segment of the series of ratios. We define it as the sum of μ and ν. We may state the definition more shortly as follows:—

The *arithmetical sum of two real numbers* is the class of the arithmetical sums of a member of the one and a member of the other chosen in all possible ways.

We can define the arithmetical product of two real numbers in exactly the same way, by multiplying a member of the one by a member of the other in all possible ways. The class of ratios thus generated is defined as the product of the two real numbers. (In all such definitions, the series of ratios is to be defined as excluding 0 and infinity.)

There is no difficulty in extending our definitions to positive and negative real numbers and their addition and multiplication.

It remains to give the definition of complex numbers.

Complex numbers, though capable of a geometrical interpretation, are not demanded by geometry in the same imperative way in which irrationals are demanded. A "complex" number means a number involving the square root of a negative number, whether integral, fractional, or real. Since the square of a negative number is positive, a number whose square is to be negative has to be a new sort of number. Using the letter i for the square root of -1, any number involving the square root of a negative number can be expressed in the form $x+yi$, where x and y are real. The part yi is called the "imaginary" part of this number, x being the "real" part. (The reason for the phrase "real numbers" is that they are contrasted with such as are "imaginary.") Complex numbers have been for a long time habitually used by mathematicians, in spite of the absence of any precise definition. It has been simply assumed that they would obey the usual arithmetical rules, and on this assumption their employment has been found profitable. They are required less for geometry than for algebra and analysis. We desire, for example, to be able to say that every quadratic equation has two roots, and every cubic equation has three, and so on. But if we are confined to real numbers, such an equation as $x^2+1=0$ has no roots, and such an equation as $x^3-1=0$ has only one. Every generalization of number has first presented itself as needed for some simple problem: negative numbers were needed in order that subtraction might be always possible, since otherwise $a-b$ would be meaningless if a were less than b; fractions were needed in order that division might be always possible; and complex numbers are needed in order that extraction of roots and solution of equations may be always possible. But extensions of number are not *created* by the mere need for them: they are created by the definition, and it is to the definition of complex numbers that we must now turn our attention.

A complex number may be regarded and defined as simply an ordered couple of real numbers. Here, as elsewhere, many definitions are possible. All that is necessary is that the definitions adopted shall lead to certain properties. In the case of complex numbers, if they are defined as ordered couples of real numbers, we secure at once some of the properties required, namely, that two real numbers are required to determine a complex number, and that among these we can distinguish a first and a second, and that two complex numbers are only identical when the first real number involved in the one is equal to the first involved in the other, and the second to the second. What is needed further can be secured by defining the rules of addition and multiplication. We are to have

$$(x+yi)+(x'+y'i) = (x+x')+(y+y')i$$
$$(x+yi)(x'+y'i) = (xx'-yy')+(xy'+x'y)i.$$

Thus we shall define that, given two ordered couples of real numbers, (x, y) and (x', y'), their sum is to be the couple $(x+x', y+y')$, and their product is to be the couple $(xx'-yy', xy'+x'y)$. By these definitions we shall secure that our ordered couples shall have the properties we desire. For example, take the product of the two couples $(0, y)$ and $(0, y')$. This will, by the above rule, be the couple $(-yy', 0)$. Thus the square of the couple (0, 1) will be the couple (−1, 0). Now those couples in which the second term is 0 are those which, according to the usual nomenclature, have their imaginary part zero; in the notation $x+yi$, they are $x+0i$, which it is natural to write simply x. Just as it is natural (but erroneous) to identify ratios whose denominator is unity with integers, so it is natural (but erroneous) to identify complex numbers whose imaginary part is zero with real numbers. Although this is an error in theory, it is a convenience in practice; "$x+0$i" may be replaced simply by "x" and "$0+yi$" by "yi," provided we remember that the "x" is not really a real number, but a special case of a complex number. And when y is 1, "yi" may of course be replaced by "i." Thus the couple (0, 1) is represented by i, and the couple (−1, 0) is represented by −1. Now our rules of multiplication make the square of (0, 1) equal to (−1, 0), *i.e.* the square of i is −1. This is what we desired to secure. Thus our definitions serve all necessary purposes.

It is easy to give a geometrical interpretation of complex numbers in the geometry of the plane. This subject was agreeably expounded by W. K. Clifford in his *Common Sense of the Exact Sciences*, a book of great merit, but written before the importance of purely logical definitions had been realized.

Complex numbers of a higher order, though much less useful and important than those what we have been defining, have certain uses that are not without importance in geometry, as may be seen, for example, in Dr Whitehead's *Universal Algebra.* The definition of complex numbers of order n is obtained by an obvious extension of the definition we have given. We define a complex number of order n as a one-many relation whose domain consists of certain real numbers and whose converse domain consists of the integers from 1 to n. [19] This is what would ordinarily be indicated by the notation $(x_1, x_2, x_3, \ldots x_n)$, where the suffixes denote correlation with the integers used as suffixes, and the correlation is one-many, not necessarily one-one, because x_r and x_s may be equal when r and s are not equal. The above definition, with a suitable rule of multiplication, will serve all purposes for which complex numbers of higher orders are needed.

[19] Cf. *Principles of Mathematics*, §360, p. 379.

We have now completed our review of those extensions of number which do not involve infinity. The application of number to infinite collections must be our next topic.

CHAPTER VIII

INFINITE CARDINAL NUMBERS

The definition of cardinal numbers which we gave in Chapter II. was applied in Chapter III. to finite numbers, *i.e.* to the ordinary natural numbers. To these we gave the name "inductive numbers," because we found that they are to be defined as numbers which obey mathematical induction starting from 0. But we have not yet considered collections which do not have an inductive number of terms, nor have we inquired whether such collections can be said to have a number at all. This is an ancient problem, which has been solved in our own day, chiefly by Georg Cantor. In the present chapter we shall attempt to explain the theory of transfinite or infinite cardinal numbers as it results from a combination of his discoveries with those of Frege on the logical theory of numbers.

It cannot be said to be *certain* that there are in fact any infinite collections in the world. The assumption that there are is what we call the "axiom of infinity." Although various ways suggest themselves by which we might hope to prove this axiom, there is reason to fear that they are all fallacious, and that there is no conclusive logical reason for believing it to be true. At the same time, there is certainly no logical reason *against* infinite collections, and we are therefore justified, in logic, in investigating the hypothesis that there are such collections. The practical form of this hypothesis, for our present purposes, is the assumption that, if n is any inductive number, n is not equal to $n+1$. Various subtleties arise in identifying this form of our assumption with the form that asserts the existence of infinite collections; but we will leave these out of account until, in a later chapter, we come to consider the axiom of infinity on its own account. For the present we shall merely assume that, if n is an inductive number, n is not equal to $n+1$. This is involved in Peano's assumption that no two inductive numbers have the same successor; for, if $n=n+1$, then $n-1$ and n have the same successor, namely n. Thus we are assuming nothing that was not involved in Peano's primitive propositions.

Let us now consider the collection of the inductive numbers themselves. This is a perfectly well-defined class. In the first place, a cardinal number is a set of classes which are all similar to each other and are not similar to anything except each other. We then define as the "inductive numbers" those among cardinals which belong to the posterity of 0 with respect to the relation of n to $n+1$, *i.e.* those which possess every property possessed by 0 and by the successors of possessors, meaning by the "successor" of n the number $n+1$. Thus the class of "inductive numbers" is perfectly definite. By our general definition of cardinal numbers, the number of terms in the class of inductive numbers is to be defined as "all those classes that are similar to the class of inductive numbers"—*i.e.* this set of classes is the number of the inductive numbers according to our definitions.

Now it is easy to see that this number is not one of the inductive numbers. If n is any inductive number, the number of numbers from 0 to n (both included) is $n+1$; therefore the total number of inductive numbers is greater than n, no matter which of the inductive numbers n may be. If we arrange the inductive numbers in a series in order of magnitude, this series has no last term; but if n is an inductive number, every series whose field has n terms has a last term, as it is easy to prove. Such differences might be multiplied *ad lib.* Thus the number of inductive numbers is a new number, different from all of them, not possessing all inductive properties. It may happen that 0 has a certain property, and that if

n has it so has $n+1$, and yet that this new number does not have it. The difficulties that so long delayed the theory of infinite numbers were largely due to the fact that some, at least, of the inductive properties were wrongly judged to be such as *must* belong to all numbers; indeed it was thought that they could not be denied without contradiction. The first step in understanding infinite numbers consists in realising the mistakenness of this view.

The most noteworthy and astonishing difference between an inductive number and this new number is that this new number is unchanged by adding 1 or subtracting 1 or doubling or halving or any of a number of other operations which we think of as necessarily making a number larger or smaller. The fact of being not altered by the addition of 1 is used by Cantor for the definition of what he calls "transfinite" cardinal numbers; but for various reasons, some of which will appear as we proceed, it is better to define an infinite cardinal number as one which does not possess all inductive properties, *i.e.* simply as one which is not an inductive number. Nevertheless, the property of being unchanged by the addition of 1 is a very important one, and we must dwell on it for a time.

To say that a class has a number which is not altered by the addition of 1 is the same thing as to say that, if we take a term x which does not belong to the class, we can find a one-one relation whose domain is the class and whose converse domain is obtained by adding x to the class. For in that case, the class is similar to the sum of itself and the term x, *i.e.* to a class having one extra term; so that it has the same number as a class with one extra term, so that if n is this number, $n=n+1$. In this case, we shall also have $n=n-1$, *i.e.* there will be one-one relations whose domains consist of the whole class and whose converse domains consist of just one term short of the whole class. It can be shown that the cases in which this happens are the same as the apparently more general cases in which *some* part (short of the whole) can be put into one-one relation with the whole. When this can be done, the correlator by which it is done may be said to "reflect" the whole class into a part of itself; for this reason, such classes will be called "reflexive." Thus:

A "reflexive" class is one which is similar to a proper part of itself. (A "proper part" is a part short of the whole.)

A "reflexive" cardinal number is the cardinal number of a reflexive class.

We have now to consider this property of reflexiveness.

One of the most striking instances of a "reflexion" is Royce's illustration of the map: he imagines it decided to make a map of England upon a part of the surface of England. A map, if it is accurate, has a perfect one-one correspondence with its original; thus our map, which is part, is in one-one relation with the whole, and must contain the same number of points as the whole, which must therefore be a reflexive number. Royce is interested in the fact that the map, if it is correct, must contain a map of the map, which must in turn contain a map of the map of the map, and so on *ad infinitum*. This point is interesting, but need not occupy us at this moment. In fact, we shall do well to pass from picturesque illustrations to such as are more completely definite, and for this purpose we cannot do better than consider the number-series itself.

The relation of n to $n+1$, confined to inductive numbers, is one-one, has the whole of the inductive numbers for its domain, and all except 0 for its converse domain. Thus the whole class of inductive numbers is similar to what the same class becomes when we omit 0. Consequently it is a "reflexive" class according to the definition, and the number of its terms is a "reflexive" number. Again, the relation of n to $2n$, confined to inductive numbers, is one-one, has the whole of the inductive numbers for its domain, and the even inductive numbers alone for its converse domain. Hence the total number of inductive numbers is the same as the number of even inductive numbers. This property was used by Leibniz (and many others) as a proof that infinite numbers are impossible; it was thought self-contradictory that "the part should be equal to the whole." But this is one of those phrases that depend for their plausibility upon an unperceived vagueness: the word "equal" has many meanings, but if it is taken to mean what we have called "similar," there is no contradiction, since an infinite collection can perfectly well have parts similar to itself. Those who regard this as impossible have, unconsciously as a rule, attributed to numbers in general properties which can only be proved by mathematical induction, and which only their familiarity makes us regard, mistakenly, as true beyond the region of the finite.

Whenever we can "reflect" a class into a part of itself, the same relation will necessarily reflect that part into a smaller part, and so on *ad infinitum*. For example, we can reflect, as we have just seen, all the inductive numbers into the even numbers; we can, by the same relation (that of n to $2n$) reflect the even numbers into the multiples of 4, these into the multiples of 8, and so on. This is an abstract analogue to Royce's problem of the map. The even numbers are a "map" of all the inductive numbers; the multiples of 4 are a map of the map; the multiples of 8 are a map of the map of the map; and so on. If we had applied the same process to the relation of n to $n+1$, our "map" would have consisted of all the inductive numbers except 0; the map of the map would have consisted of all from 2 onward, the map of the map of the map of all from 3 onward; and so on. The chief use of such illustrations is in order to become familiar with the idea of reflexive classes, so that apparently paradoxical arithmetical propositions can be readily translated into the language of reflections and classes, in which the air of paradox is much less.

It will be useful to give a definition of the number which is that of the inductive cardinals. For this purpose we will first define the kind of series exemplified by the inductive cardinals in order of magnitude. The kind of series which is called a "progression" has already been considered in Chapter I. It is a series which can be generated by a relation of consecutiveness: every member of the series is to have a successor, but there is to be just one which has no predecessor, and every member of the series is to be in the posterity of this term with respect to the relation "immediate predecessor." These characteristics may be summed up in the following definition:—[20]

[20] Cf. *Principia Mathematica*, vol. ii. *123.

A "progression" is a one-one relation such that there is just one term belonging to the domain but not to the converse domain, and the domain is identical with the posterity of this one term.

It is easy to see that a progression, so defined, satisfies Peano's five axioms. The term belonging to the domain but not to the converse domain will be what he calls "0"; the term to which a term has the one-one relation will be the "successor" of the term; and the domain of the one-one relation will be what he calls "number." Taking his five axioms in turn, we have the following translations:—

(1) "0 is a number" becomes: "The member of the domain which is not a member of the converse domain is a member of the domain." This is equivalent to the existence of such a member, which is given in our definition. We will call this member "the first term."

(2) "The successor of any number is a number" becomes: "The term to which a given member of the domain has the relation in question is again a member of the domain." This is proved as follows: By the definition, every member of the domain is a member of the posterity of the first term; hence the successor of a member of the domain must be a member of the posterity of the first term (because the posterity of a term always contains its own successors, by the general definition of posterity), and therefore a member of the domain, because by the definition the posterity of the first term is the same as the domain.

(3) "No two numbers have the same successor." This is only to say that the relation is one-many, which it is by definition (being one-one).

(4) "0 is not the successor of any number" becomes: "The first term is not a member of the converse domain," which is again an immediate result of the definition.

(5) This is mathematical induction, and becomes: "Every member of the domain belongs to the posterity of the first term," which was part of our definition.

Thus progressions as we have defined them have the five formal properties from which Peano deduces arithmetic. It is easy to show that two progressions are "similar" in the sense defined for similarity of relations in Chapter VI. We can, of course, derive a relation which is serial from the one-one relation by which we define a progression: the method used is that explained in Chapter IV., and the relation is that of a term to a member of its proper posterity with respect to the original one-one relation.

Two transitive asymmetrical relations which generate progressions are similar, for the same reasons for which the corresponding one-one relations are similar. The class of all such transitive generators of progressions is a "serial number" in the sense of Chapter VI.; it is in fact the smallest of infinite serial numbers, the number to which Cantor has given the name ω, by which he has made it famous.

But we are concerned, for the moment, with *cardinal* numbers. Since two progressions are similar relations, it follows that their domains (or their fields, which are the same as their domains) are similar classes. The domains of progressions form a cardinal number, since every class which is similar to the domain of a progression is easily shown to be itself the domain of a progression. This cardinal number is the smallest of the infinite cardinal numbers; it is the one to which Cantor has appropriated the Hebrew Aleph with the suffix 0, to distinguish it from larger infinite cardinals, which have other suffixes. Thus the name of the smallest of infinite cardinals is $\aleph_0$.

To say that a class has $\aleph_0$ terms is the same thing as to say that it is a member of $\aleph_0$, and this is the same thing as to say that the members of the class can be arranged in a progression. It is obvious that any progression remains a progression if we omit a finite number of terms from it, or every other term, or all except every tenth term or every hundredth term. These methods of thinning out a progression do not make it cease to be a progression, and therefore do not diminish the number of its terms, which remains ℵ0. In fact, any selection from a progression is a progression if it has no last term, however sparsely it may be distributed. Take (say) inductive numbers of the form n^n, or n^{nn}. Such numbers grow very rare in the higher parts of the number series, and yet there are just as many of them as there are inductive numbers altogether, namely, $\aleph_0$.

Conversely, we can add terms to the inductive numbers without increasing their number. Take, for example, ratios. One might be inclined to think that there must be many more ratios than integers, since ratios whose denominator is 1 correspond to the integers, and seem to be only an infinitesimal proportion of ratios. But in actual fact the number of ratios (or fractions) is exactly the same as the number of inductive numbers, namely, $\aleph_0$. This is easily seen by arranging ratios in a series on the following plan: If the sum of numerator and denominator in one is less than in the other, put the one before the other; if the sum is equal in the two, put first the one with the smaller numerator. This gives us the series

1, 1/2, 2, 1/3, 3, 1/4, 2/3, 3/2, 4, 1/5, ...

This series is a progression, and all ratios occur in it sooner or later. Hence we can arrange all ratios in a progression, and their number is therefore $\aleph_0$.

It is not the case, however, that *all* infinite collections have $\aleph_0$ terms. The number of real numbers, for example, is greater than $\aleph_0$; it is, in fact, $2\aleph_0$, and it is not hard to prove that 2^n is greater than n even when n is infinite. The easiest way of proving this is to prove, first, that if a class has n members, it contains 2^n sub-classes—in other words, that there are 2^n ways of selecting some of its members (including the extreme cases where we select all or none); and secondly, that the number of sub-classes contained in a class is always greater than the number of members of the class. Of these two propositions, the first is familiar in the case of finite numbers, and is not hard to extend to infinite numbers. The proof of the second is so simple and so instructive that we shall give it:

In the first place, it is clear that the number of sub-classes of a given class (say α) is at least as great as the number of members, since each member constitutes a sub-class, and we thus have a correlation of all the members with some of the sub-classes. Hence it follows that, if the number of sub-classes is not *equal* to the number of members, it must be *greater*. Now it is easy to prove that the number is not equal, by showing that, given any one-one relation whose domain is the members and whose converse domain is contained among the set of sub-classes, there must be at least one sub-class not belonging to the converse domain. The proof is as follows: [21] When a one-one correlation R is established between all the members of α and some of the sub-classes, it may happen that a given member x is correlated with a sub-class of which it is a member; or, again, it may happen that x is correlated with a sub-class of which it is not a member. Let us form the whole class, β say, of those members x which are correlated with sub-classes of which they are not members. This is a sub-class of α, and it is not correlated with any member of α. For, taking first the members of β, each of them is (by the definition of β) correlated with some sub-class of which it is not a member, and is therefore not correlated with β. Taking next the terms which are not members of β, each of them (by the definition of β) is correlated with some sub-class of which it is a member, and therefore again is not correlated with β. Thus no member of α is correlated with β. Since R was *any* one-one correlation of all members with some sub-classes, it follows that there is no correlation of all members with *all* sub-classes. It does not matter to the proof if β has no members: all that happens in that case is that the sub-class which is shown to be omitted is the null-class. Hence in any case the number of sub-classes is not equal to the number of members, and therefore, by what was said earlier, it is greater. Combining this with the proposition that, if n is the number of members, 2^n is the number of sub-classes, we have the theorem that 2^n is always greater than n, even when n is infinite.

[21] This proof is taken from Cantor, with some simplifications: see *Jahresbericht der Deutschen Mathematiker-Vereinigung*, i. (1892), p. 77.

It follows from this proposition that there is no maximum to the infinite cardinal numbers. However great an infinite number n may be, 2^n will be still greater. The arithmetic of infinite numbers is somewhat surprising until one becomes accustomed to it. We have, for example,

$$\aleph_0+1 = \aleph_0,$$
$$\aleph_0+n = \aleph_0, \text{ where } n \text{ is any inductive number,}$$
$$\aleph_0^2 = \aleph_0.$$

(This follows from the case of the ratios, for, since a ratio is determined by a pair of inductive numbers, it is easy to see that the number of ratios is the square of the number of inductive numbers, *i.e.* it is $\aleph_0^2$; but we saw that it is also $\aleph_0$.)

$$\aleph_0 n = \aleph_0, \text{ where } n \text{ is any inductive number.}$$

(This follows from $\aleph_0^2 = \aleph_0$ by induction; for if $\aleph_0^n = \aleph_0$,
then $\aleph_0^{n+1} = \aleph_0^2 = \aleph_0$.)
But $2^{\aleph_0} > \aleph_0$.

In fact, as we shall see later, $2^{\aleph_0}$ is a very important number, namely, the number of terms in a series which has "continuity" in the sense in which this word is used by Cantor. Assuming space and time to be continuous in this sense (as we commonly do in analytical geometry and kinematics), this will be the number of points in space or of instants in time; it will also be the number of points in any finite portion of space, whether line, area, or volume. After $\aleph_0$, $2^{\aleph_0}$ is the most important and interesting of infinite cardinal numbers.

Although addition and multiplication are always possible with infinite cardinals, subtraction and division no longer give definite results, and cannot therefore be employed as they are employed in elementary arithmetic. Take subtraction to begin with: so long as the number subtracted is finite, all goes well; if the other number is reflexive, it remains unchanged. Thus $\aleph_0 - n = \aleph_0$, if n is finite; so far, subtraction gives a perfectly definite result. But it is otherwise when we subtract $\aleph_0$ from itself; we may then get any result, from 0 up to $\aleph_0$. This is easily seen by examples. From the inductive numbers, take away the following collections of $\aleph_0$ terms:—

(1) All the inductive numbers—remainder, zero.

(2) All the inductive numbers from n onwards—remainder, the numbers from 0 to $n-1$, numbering n terms in all.

(3) All the odd numbers—remainder, all the even numbers, numbering $\aleph_0$ terms.

All these are different ways of subtracting $\aleph_0$ from $\aleph_0$, and all give different results.

As regards division, very similar results follow from the fact that $\aleph_0$ is unchanged when multiplied by 2 or 3 or any finite number n or by $\aleph_0$. It follows that $\aleph_0$ divided by $\aleph 0$ may have any value from 1 up to $\aleph_0$.

From the ambiguity of subtraction and division it results that negative numbers and ratios cannot be extended to infinite numbers. Addition, multiplication, and exponentiation proceed quite satisfactorily, but the inverse operations—subtraction, division, and extraction of roots—are ambiguous, and the notions that depend upon them fail when infinite numbers are concerned.

The characteristic by which we defined finitude was mathematical induction, *i.e.* we defined a number as finite when it obeys mathematical induction starting from 0, and a class as finite when its number is finite. This definition yields the sort of result that a definition ought to yield, namely, that the finite numbers are those that occur in the ordinary number-series 0, 1, 2, 3, ... But in the present chapter, the infinite numbers we have discussed have not merely been non-inductive: they have also been *reflexive*. Cantor used reflexiveness as the *definition* of the infinite, and believes that it is equivalent to non-inductiveness; that is to say, he believes that every class and every cardinal is either inductive or reflexive. This may be true, and may very possibly be capable of proof; but the proofs hitherto offered by Cantor and others (including the present author in former days) are fallacious, for reasons which will be explained when we come to consider the "multiplicative axiom." At present, it is not known whether there are classes and cardinals which are neither reflexive nor inductive. If n were such a cardinal, we should not have $n=n+1$, but n would not be one of the "natural numbers," and would be lacking in some of the inductive properties. All *known* infinite classes and cardinals are reflexive; but for the present it is well to preserve an open mind as to whether there are instances, hitherto unknown, of classes and cardinals which are neither reflexive nor inductive. Meanwhile, we adopt the following definitions:—

A *finite* class or cardinal is one which is *inductive*.

An *infinite* class or cardinal is one which is *not inductive*. All *reflexive* classes and cardinals are infinite; but it is not known at present whether all infinite classes and cardinals are reflexive. We shall return to this subject in Chapter XII.

CHAPTER IX

INFINITE SERIES AND ORDINALS

An "infinite series" may be defined as a series of which the field is an infinite class. We have already had occasion to consider one kind of infinite series, namely, progressions. In this chapter we shall consider the subject more generally.

The most noteworthy characteristic of an infinite series is that its serial number can be altered by merely re-arranging its terms. In this respect there is a certain oppositeness between cardinal and serial numbers. It is possible to keep the cardinal number of a reflexive class unchanged in spite of adding terms to it; on the other hand, it is possible to change the serial number of a series without adding or taking away any terms, by mere re-arrangement. At the same time, in the case of any infinite series it is also possible, as with cardinals, to add terms without altering the serial number: everything depends upon the way in which they are added.

In order to make matters clear, it will be best to begin with examples. Let us first consider various different kinds of series which can be made out of the inductive numbers arranged on various plans. We start with the series

$$1, 2, 3, 4, \ldots n, \ldots,$$

which, as we have already seen, represents the smallest of infinite serial numbers, the sort that Cantor calls ω. Let us proceed to thin out this series by repeatedly performing the operation of removing to the end the first even number that occurs. We thus obtain in succession the various series:

$$1, 3, 4, 5, \ldots n, \ldots 2,$$
$$1, 3, 5, 6, \ldots n+1, \ldots 2, 4,$$
$$1, 3, 5, 7, \ldots n+2, \ldots 2, 4, 6,$$

and so on. If we imagine this process carried on as long as possible, we finally reach the series

$$1, 3, 5, 7, \dots 2n+1, \dots 2, 4, 6, 8, \dots 2n, \dots,$$

in which we have first all the odd numbers and then all the even numbers.

The serial numbers of these various series are $\omega+1$, $\omega+2$, $\omega+3$, ... 2ω. Each of these numbers is "greater" than any of its predecessors, in the following sense:—

One serial number is said to be "greater" than another if any series having the first number contains a part having the second number, but no series having the second number contains a part having the first number.

If we compare the two series

$$1, 2, 3, 4, \dots n, \dots$$
$$1, 3, 4, 5, \dots n+1, \dots 2,$$

we see that the first is similar to the part of the second which omits the last term, namely, the number 2, but the second is not similar to any part of the first. (This is obvious, but is easily demonstrated.) Thus the second series has a greater serial number than the first, according to the definition—*i.e.* $\omega+1$ is greater than ω. But if we add a term at the beginning of a progression instead of the end, we still have a progression. Thus $1+\omega=\omega$. Thus $1+\omega$ is not equal to $\omega+1$. This is characteristic of relation-arithmetic generally: if μ and ν are two relation-numbers, the general rule is that $\mu+\nu$ is not equal to $\nu+\mu$. The case of finite ordinals, in which there is equality, is quite exceptional.

The series we finally reached just now consisted of first all the odd numbers and then all the even numbers, and its serial number is 2ω. This number is greater than ω or $\omega+n$, where n is finite. It is to be observed that, in accordance with the general definition of order, each of these arrangements of integers is to be regarded as resulting from some definite relation. *E.g.* the one which merely removes 2 to the end will be defined by the following relation: "x and y are finite integers, and either y is 2 and x is not 2, or neither is 2 and x is less than y." The one which puts first all the odd numbers and then all the even ones will be defined by: "x and y are finite integers, and either x is odd and y is even or x is less than y and both are odd or both are even." We shall not trouble, as a rule, to give these formulæ in future; but the fact that they *could* be given is essential.

The number which we have called 2ω, namely, the number of a series consisting of two progressions, is sometimes called $\omega.2$. Multiplication, like addition, depends upon the order of the factors: a progression of couples gives a series such as

$$x_1, y_1, x_2, y_2, x_3, y_3, \dots x_n, y_n, \dots,$$

which is itself a progression; but a couple of progressions gives a series which is twice as long as a progression. It is therefore necessary to distinguish between 2ω and $\omega.2$. Usage is variable; we shall use 2ω for a couple of progressions and $\omega.2$ for a progression of couples, and this decision of course governs our general interpretation of "$\alpha.\ \beta$" when α and β are relation-numbers: "$\alpha.\ \beta$" will have to stand for a suitably constructed sum of α relations each having β terms.

We can proceed indefinitely with the process of thinning out the inductive numbers. For example, we can place first the odd numbers, then their doubles, then the doubles of these, and so on. We thus obtain the series

$$1, 3, 5, 7, \ldots ; 2, 6, 10, 14, \ldots ; 4, 12, 20, 28, \ldots ; 8, 24, 40, 56, \ldots ,$$

of which the number is ω^2, since it is a progression of progressions. Any one of the progressions in this new series can of course be thinned out as we thinned out our original progression. We can proceed to ω^3, ω^4, ... ω^ω, and so on; however far we have gone, we can always go further.

The series of all the ordinals that can be obtained in this way, *i.e.* all that can be obtained by thinning out a progression, is itself longer than any series that can be obtained by re-arranging the terms of a progression. (This is not difficult to prove.) The cardinal number of the class of such ordinals can be shown to be greater than $\aleph_0$; it is the number which Cantor calls $\aleph_1$. The ordinal number of the series of all ordinals that can be made out of an $\aleph_0$, taken in order of magnitude, is called ω_1. Thus a series whose ordinal number is ω_1 has a field whose cardinal number is $\aleph_1$.

We can proceed from $\omega 1$ and $\aleph_1$ to ω_2 and $\aleph_2$ by a process exactly analogous to that by which we advanced from ω and $\aleph_0$ to ω_1 and $\aleph_1$. And there is nothing to prevent us from advancing indefinitely in this way to new cardinals and new ordinals. It is not known whether $2^{\aleph_0}$ is equal to any of the cardinals in the series of Alephs. It is not even known whether it is comparable with them in magnitude; for aught we know, it may be neither equal to nor greater nor less than any one of the Alephs. This question is connected with the multiplicative axiom, of which we shall treat later.

All the series we have been considering so far in this chapter have been what is called "well-ordered." A well-ordered series is one which has a beginning, and has consecutive terms, and has a term next after any selection of its terms, provided there are any terms after the selection. This excludes, on the one hand, compact series, in which there are terms between any two, and on the other hand series which have no beginning, or in which there are subordinate parts having no beginning. The series of negative integers in order of magnitude, having no beginning, but ending with -1, is not well-ordered; but taken in the reverse order, beginning with -1, it is well-ordered, being in fact a progression. The definition is:

A "well-ordered" series is one in which every sub-class (except, of course, the null-class) has a first term.

An "ordinal" number means the relation-number of a well-ordered series. It is thus a species of serial number.

Among well-ordered series, a generalized form of mathematical induction applies. A property may be said to be "transfinitely hereditary" if, when it belongs to a certain selection of the terms in a series, it belongs to their immediate successor provided they have one. In a well-ordered series, a transfinitely hereditary property belonging to the first term of the series belongs to the whole series. This makes it possible to prove many propositions concerning well-ordered series which are not true of all series.

It is easy to arrange the inductive numbers in series which are not well-ordered, and even to arrange them in compact series. For example, we can adopt the following plan: consider the decimals from ·1 (inclusive) to 1 (exclusive), arranged in order of magnitude. These form a compact series; between any two there are always an infinite number of others. Now omit the dot at the beginning of each, and we have a compact series consisting of all finite integers except such as divide by 10. If we wish to include those that divide by 10, there is no difficulty; instead of starting with ·1, we will include all decimals less than 1, but when we remove the dot, we will transfer to the right any 0'*s* that occur at the beginning of our decimal. Omitting these, and returning to the ones that have no 0'*s* at the beginning, we can state the rule for the arrangement of our integers as follows: Of two integers that do not begin with the same digit, the one that begins with the smaller digit comes first. Of two that do begin with the same digit, but differ at the second digit, the one with the smaller second digit comes first, but first of all the one with no second digit; and so on. Generally, if two integers agree as regards the first n digits, but not as regards the $(n+1)^{th}$, that one comes first which has either no $(n+1)^{th}$ digit or a smaller one than the other. This rule of arrangement, as the reader can easily convince himself, gives rise to a compact series containing all the integers not divisible by 10; and, as we saw, there is no difficulty about including those that are divisible by 10. It follows from this example that it is possible to construct compact series having $\aleph_0$ terms. In fact, we have already seen that there are $\aleph_0$ ratios, and ratios in order of magnitude form a compact series; thus we have here another example. We shall resume this topic in the next chapter.

Of the usual formal laws of addition, multiplication, and exponentiation, all are obeyed by transfinite cardinals, but only some are obeyed by transfinite ordinals, and those that are obeyed by them are obeyed by all relation-numbers. By the "usual formal laws" we mean the following:—

I. The commutative law:
$\alpha+\beta=\beta+\alpha$ and $\alpha\times\beta=\beta\times\alpha$.
II. The associative law:
$(\alpha+\beta)+\gamma=\alpha+(\beta+\gamma)$ and $(\alpha\times\beta)\times\gamma=\alpha\times(\beta\times\gamma)$.
III. The distributive law:
$\alpha(\beta+\gamma)=\alpha\beta+\alpha\gamma$.

When the commutative law does not hold, the above form of the distributive law must be distinguished from

$$(\beta+\gamma)\alpha=\beta\alpha+\gamma\alpha.$$

As we shall see immediately, one form may be true and the other false.

IV. The laws of exponentiation:
$\alpha^\beta.\alpha^\gamma=\alpha^{\beta+\gamma}$, $\alpha^\gamma.\ \beta^\gamma=(\alpha\beta)^\gamma$, $(\alpha^\beta)^\gamma=\alpha^{\beta\gamma}$.

All these laws hold for cardinals, whether finite or infinite, and for *finite* ordinals. But when we come to infinite ordinals, or indeed to relation-numbers in general, some hold and some do not. The commutative law does not hold; the associative law does hold; the distributive law (adopting the convention we have adopted above as regards the order of the factors in a product) holds in the form

$$(\beta+\gamma)\alpha=\beta\alpha+\gamma\alpha,$$

but not in the form

$$\alpha(\beta+\gamma)=\alpha\beta+\alpha\gamma;$$

the exponential laws

$$\alpha+.\ \alpha^{\gamma}=\alpha^{\beta+\gamma} \text{ and } (\alpha^{\beta})^{\gamma}=\alpha^{\beta\gamma}$$

still hold, but not the law

$$\alpha^{\gamma}.\ \beta^{\gamma}=(\alpha\beta)^{\gamma},$$

which is obviously connected with the commutative law for multiplication.

The definitions of multiplication and exponentiation that are assumed in the above propositions are somewhat complicated. The reader who wishes to know what they are and how the above laws are proved must consult the second volume of *Principia Mathematica*, *172–176.

Ordinal transfinite arithmetic was developed by Cantor at an earlier stage than cardinal transfinite arithmetic, because it has various technical mathematical uses which led him to it. But from the point of view of the philosophy of mathematics it is less important and less fundamental than the theory of transfinite cardinals. Cardinals are essentially simpler than ordinals, and it is a curious historical accident that they first appeared as an abstraction from the latter, and only gradually came to be studied on their own account. This does not apply to Frege's work, in which cardinals, finite and transfinite, were treated in complete independence of ordinals; but it was Cantor's work that made the world aware of the subject, while Frege's remained almost unknown, probably in the main on account of the difficulty of his symbolism. And mathematicians, like other people, have more difficulty in understanding and using notions which are comparatively "simple" in the logical sense than in manipulating more complex notions which are more akin to their ordinary practice. For these reasons, it was only gradually that the true importance of cardinals in mathematical philosophy was recognised. The importance of ordinals, though by no means small, is distinctly less than that of cardinals, and is very largely merged in that of the more general conception of relation-numbers.

CHAPTER X

LIMITS AND CONTINUITY

The conception of a "limit" is one of which the importance in mathematics has been found continually greater than had been thought. The whole of the differential and integral calculus, indeed practically everything in higher mathematics, depends upon limits. Formerly, it was supposed that infinitesimals were involved in the foundations of these subjects, but Weierstrass showed that this is an error: wherever infinitesimals were thought to occur, what really occurs is a set of finite quantities having zero for their lower limit. It used to be thought that "limit" was an essentially quantitative notion, namely, the notion of a quantity to which others approached nearer and nearer, so that among those others there would be some differing by less than any assigned quantity. But in fact the notion of "limit" is a purely ordinal notion, not involving quantity at all (except by accident when the series concerned happens to be quantitative). A given point on a line may be the limit of a set of points on the line, without its being necessary to bring in co-ordinates or measurement or anything quantitative. The cardinal number $\aleph_0$ is the limit (in the order of magnitude) of the cardinal numbers 1, 2, 3, ... *n*, ..., although the numerical difference between $\aleph_0$ and a finite cardinal is constant and infinite: from a quantitative point of view, finite numbers get no nearer to $\aleph_0$ as they grow larger. What makes $\aleph_0$ the limit of the finite numbers is the fact that, in the series, it comes immediately after them, which is an *ordinal* fact, not a quantitative fact.

There are various forms of the notion of "limit," of increasing complexity. The simplest and most fundamental form, from which the rest are derived, has been already defined, but we will here repeat the definitions which lead to it, in a general form in which they do not demand that the relation concerned shall be serial. The definitions are as follows:—

The "minima" of a class α with respect to a relation P are those members of α and the field of P (if any) to which no member of α has the relation P.

The "maxima" with respect to P are the minima with respect to the converse of P.

The "sequents" of a class α with respect to a relation P are the minima of the "successors" of α, and the "successors" of α are those members of the field of P to which every member of the common part of α and the field of P has the relation P.

The "precedents" with respect to P are the sequents with respect to the converse of P.

The "upper limits" of α with respect to P are the sequents provided α has no maximum; but if α has a maximum, it has no upper limits.

The "lower limits" with respect to P are the upper limits with respect to the converse of P.

Whenever P has connexity, a class can have at most one maximum, one minimum, one sequent, etc. Thus, in the cases we are concerned with in practice, we can speak of "*the* limit" (if any).

When P is a serial relation, we can greatly simplify the above definition of a limit. We can, in that case, define first the "boundary" of a class α, *i.e.* its limit or maximum, and then proceed to distinguish the case where the boundary is the limit from the case where it is a maximum. For this purpose it is best to use the notion of "segment."

We will speak of the "segment of P defined by a class α" as all those terms that have the relation P to some one or more of the members of α. This will be a segment in the sense defined in Chapter VII.; indeed, every segment in the sense there defined is the segment defined by some class α. If P is serial, the segment defined by α consists of all the terms that precede some term or other of α. If α has a maximum, the segment will be all the predecessors of the maximum. But if α has no maximum, every member of α precedes some other member of α, and the whole of α is therefore included in the segment defined by α. Take, for example, the class consisting of the fractions

$$\tfrac{1}{2}, \tfrac{3}{4}, \tfrac{7}{8}, \tfrac{15}{16}, \ldots,$$

i.e. of all fractions of the form $1-\tfrac{1}{2}^n$ for different finite values of *n*. This series of fractions has no maximum, and it is clear that the segment which it defines (in the whole series of fractions in order of magnitude) is the class of all proper fractions. Or, again, consider the prime numbers, considered as a selection from the cardinals (finite and infinite) in order of magnitude. In this case the segment defined consists of all finite integers.

Assuming that P is serial, the "boundary" of a class α will be the term *x* (if it exists) whose predecessors are the segment defined by α.

A "maximum" of α is a boundary which is a member of α.

An "upper limit" of α is a boundary which is not a member of α.

If a class has no boundary, it has neither maximum nor limit. This is the case of an "irrational" Dedekind cut, or of what is called a "gap."

Thus the "upper limit" of a set of terms α with respect to a series P is that term *x* (if it exists) which comes after all the α'*s*, but is such that every earlier term comes before some of the α'*s*.

We may define all the "upper limiting-points" of a set of terms β as all those that are the upper limits of sets of terms chosen out of β. We shall, of course, have to distinguish upper limiting-points from lower limiting-points. If we consider, for example, the series of ordinal numbers:

$$1, 2, 3, \ldots \omega, \omega+1, \ldots 2\omega, 2\omega+1, \ldots 3\omega, \ldots \omega^2, \ldots \omega^3, \ldots,$$

the upper limiting-points of the field of this series are those that have no immediate predecessors, *i.e.*

$$1, \omega, 2\omega, 3\omega, \ldots \omega^2, \omega^2+\omega, \ldots 2\omega^2, \ldots \omega+ \ldots$$

The upper limiting-points of the field of this new series will be

$$1, \omega^2, 2\omega^2, \ldots \omega^3, \omega^3+\omega^2 \ldots$$

On the other hand, the series of ordinals—and indeed every well-ordered series—has no lower limiting-points, because there are no terms except the last that have no immediate successors. But if we consider such a series as the series of ratios, every member of this series is both an upper and a lower limiting-point for suitably chosen sets. If we consider the series of real numbers, and select out of it the rational real numbers, this set (the rationals) will have all the real numbers as upper and lower limiting-points. The limiting-points of a set are called its "first derivative," and the limiting-points of the first derivative are called the second derivative, and so on.

With regard to limits, we may distinguish various grades of what may be called "continuity" in a series. The word "continuity" had been used for a long time, but had remained without any precise definition until the time of Dedekind and Cantor. Each of these two men gave a precise significance to the term, but Cantor'*s* definition is narrower than Dedekind'*s*: a series which has Cantorian continuity must have Dedekindian continuity, but the converse does not hold.

The first definition that would naturally occur to a man seeking a precise meaning for the continuity of series would be to define it as consisting in what we have called "compactness," *i.e.* in the fact that between any two terms of the series there are others. But this would be an inadequate definition, because of the existence of "gaps" in series such as the series of ratios. We saw in Chapter VII. that there are innumerable ways in which the series of ratios can be divided into two parts, of which one wholly precedes the other, and of which the first has no last term, while the second has no first term. Such a state of affairs seems contrary to the vague feeling we have as to what should characterize "continuity," and, what is more, it shows that the series of ratios is not the sort of series that is needed for many mathematical purposes. Take geometry, for example: we wish to be able to say that when two straight lines cross each other they have a point in common, but if the series of points on a line were similar to the series of ratios, the two lines might cross in a "gap" and have no point in common. This is a crude example, but many others might be given to show that compactness is inadequate as a mathematical definition of continuity.

It was the needs of geometry, as much as anything, that led to the definition of "Dedekindian" continuity. It will be remembered that we defined a series as Dedekindian when every sub-class of the field has a boundary. (It is sufficient to assume that there is always an *upper* boundary, or that there is always a *lower* boundary. If one of these is assumed, the other can be deduced.) That is to say, a series is Dedekindian when there are no gaps. The absence of gaps may arise either through terms having successors, or through the existence of limits in the absence of maxima. Thus a finite series or a well-ordered series is Dedekindian, and so is the series of real numbers. The former sort of Dedekindian series is excluded by assuming that our series is compact; in that case our series must have a property which may, for many purposes, be fittingly called continuity. Thus we are led to the definition:

A series has "Dedekindian continuity" when it is Dedekindian and compact.

But this definition is still too wide for many purposes. Suppose, for example, that we desire to be able to assign such properties to geometrical space as shall make it certain that every point can be specified by means of co-ordinates which are real numbers: this is not insured by Dedekindian continuity alone. We want to be sure that every point which cannot be specified by *rational* co-ordinates can be specified as the limit of a *progression* of points whose co-ordinates are rational, and this is a further property which our definition does not enable us to deduce.

We are thus led to a closer investigation of series with respect to limits. This investigation was made by Cantor and formed the basis of his definition of continuity, although, in its simplest form, this definition somewhat conceals the considerations which have given rise to it. We shall, therefore, first travel through some of Cantor's conceptions in this subject before giving his definition of continuity.

Cantor defines a series as "perfect" when all its points are limiting-points and all its limiting-points belong to it. But this definition does not express quite accurately what he means. There is no correction required so far as concerns the property that all its points are to be limiting-points; this is a property belonging to compact series, and to no others if all points are to be upper limiting- or all lower limiting-points. But if it is only assumed that they are limiting-points one way, without specifying which, there will be other series that will have the property in question—for example, the series of decimals in which a decimal ending in a recurring 9 is distinguished from the corresponding terminating decimal and placed immediately before it. Such a series is very nearly compact, but has exceptional terms which are consecutive, and of which the first has no immediate predecessor, while the second has no immediate successor. Apart from such series, the series in which every point is a limiting-point are compact series; and this holds without qualification if it is specified that every point is to be an upper limiting-point (or that every point is to be a lower limiting-point).

Although Cantor does not explicitly consider the matter, we must distinguish different kinds of limiting-points according to the nature of the smallest sub-series by which they can be defined. Cantor assumes that they are to be defined by progressions, or by regressions (which are the converses of progressions). When every member of our series is the limit of a progression or regression, Cantor calls our series "condensed in itself" (*insichdicht*).

We come now to the second property by which perfection was to be defined, namely, the property which Cantor calls that of being "closed" (*abgeschlossen*). This, as we saw, was first defined as consisting in the fact that all the limiting-points of a series belong to it. But this only has any effective significance if our series is *given* as contained in some other larger series (as is the case, *e.g.*, with a selection of real numbers), and limiting-points are taken in relation to the larger series. Otherwise, if a series is considered simply on its own account, it cannot fail to contain its limiting-points. What Cantor *means* is not exactly what he says; indeed, on other occasions he says something rather different, which is what he means. What he really means is that every subordinate series which is of the sort that might be expected to have a limit does have a limit within the given series; *i.e.* every subordinate series which has no maximum has a limit, *i.e.* every subordinate series has a boundary. But Cantor does not state this for *every* subordinate series, but only for progressions and regressions. (It is not clear how far he recognizes that this is a limitation.) Thus, finally, we find that the definition we want is the following:—

A series is said to be "closed" (*abgeschlossen*) when every progression or regression contained in the series has a limit in the series.

We then have the further definition:—

A series is "perfect" when it is *condensed in itself* and *closed*, *i.e.* when every term is the limit of a progression or regression, and every progression or regression contained in the series has a limit in the series.

In seeking a definition of continuity, what Cantor has in mind is the search for a definition which shall apply to the series of real numbers and to any series similar to that, but to no others. For this purpose we have to add a further property. Among the real numbers some are rational, some are irrational; although the number of irrationals is greater than the number of rationals, yet there are rationals between any two real numbers, however little the two may differ. The number of rationals, as we saw, is $\aleph_0$. This gives a further property which suffices to characterize continuity completely, namely, the property of containing a class of $\aleph_0$ members in such a way that some of this class occur between any two terms of our series, however near together. This property, added to perfection, suffices to define a class of series which are all similar and are in fact a serial number. This class Cantor defines as that of continuous series.

We may slightly simplify his definition. To begin with, we say:

A "median class" of a series is a sub-class of the field such that members of it are to be found between any two terms of the series.

Thus the rationals are a median class in the series of real numbers. It is obvious that there cannot be median classes except in compact series.

We then find that Cantor's definition is equivalent to the following:—

A series is "continuous" when (1) it is Dedekindian, (2) it contains a median class having $\aleph_0$ terms.

To avoid confusion, we shall speak of this kind as "Cantorian continuity." It will be seen that it implies Dedekindian continuity, but the converse is not the case. All series having Cantorian continuity are similar, but not all series having Dedekindian continuity.

The notions of *limit* and *continuity* which we have been defining must not be confounded with the notions of the limit of a function for approaches to a given argument, or the continuity of a function in the neighbourhood of a given argument. These are different notions, very important, but derivative from the above and more complicated. The continuity of motion (if motion is continuous) is an instance of the continuity of a function; on the other hand, the continuity of space and time (if they are continuous) is an instance of the continuity of series, or (to speak more cautiously) of a kind of continuity which can, by sufficient mathematical manipulation, be reduced to the continuity of series. In view of the fundamental importance of motion in applied mathematics, as well as for other reasons, it will be well to deal briefly with the notions of limits and continuity as applied to functions; but this subject will be best reserved for a separate chapter.

The definitions of continuity which we have been considering, namely, those of Dedekind and Cantor, do not correspond very closely to the vague idea which is associated with the word in the mind of the man in the street or the philosopher. They conceive continuity rather as absence of separateness, the sort of general obliteration of distinctions which characterizes a thick fog. A fog gives an impression of vastness without definite multiplicity or division. It is this sort of thing that a metaphysician means by "continuity," declaring it, very truly, to be characteristic of his mental life and of that of children and animals.

The general idea vaguely indicated by the word "continuity" when so employed, or by the word "flux," is one which is certainly quite different from that which we have been defining. Take, for example, the series of real numbers. Each is what it is, quite definitely and uncompromisingly; it does not pass over by imperceptible degrees into another; it is a hard, separate unit, and its distance from every other unit is finite, though it can be made less than any given finite amount assigned in advance. The question of the relation between the kind of continuity existing among the real numbers and the kind exhibited, *e.g.* by what we see at a given time, is a difficult and intricate one. It is not to be maintained that the two kinds are simply identical, but it may, I think, be very well maintained that the mathematical conception which we have been considering in this chapter gives the abstract logical scheme to which it must be possible to bring empirical material by suitable manipulation, if that material is to be called "continuous" in any precisely definable sense. It would be quite impossible to justify this thesis within the limits of the present volume. The reader who is interested may read an attempt to justify it as regards *time* in particular by the present author in the *Monist* for 1914–5, as well as in parts of *Our Knowledge of the External World.* With these indications, we must leave this problem, interesting as it is, in order to return to topics more closely connected with mathematics.

CHAPTER XI

LIMITS AND CONTINUITY OF FUNCTIONS

In this chapter we shall be concerned with the definition of the limit of a function (if any) as the argument approaches a given value, and also with the definition of what is meant by a "continuous function." Both of these ideas are somewhat technical, and would hardly demand treatment in a mere introduction to mathematical philosophy but for the fact that, especially through the so-called infinitesimal calculus, wrong views upon our present topics have become so firmly embedded in the minds of professional philosophers that a prolonged and considerable effort is required for their uprooting. It has been thought ever since the time of Leibniz that the differential and integral calculus required infinitesimal quantities. Mathematicians (especially Weierstrass) proved that this is an error; but errors incorporated, *e.g.* in what Hegel has to say about mathematics, die hard, and philosophers have tended to ignore the work of such men as Weierstrass.

Limits and continuity of functions, in works on ordinary mathematics, are defined in terms involving number. This is not essential, as Dr Whitehead has shown. [22] We will, however, begin with the definitions in the text-books, and proceed afterwards to show how these definitions can be generalized so as to apply to series in general, and not only to such as are numerical or numerically measurable.

[22] See *Principia Mathematica*, vol. ii. *230-234.

Let us consider any ordinary mathematical function *fx*, where *x* and *fx* are both real numbers, and *fx* is one-valued—*i.e.* when *x* is given, there is only one value that *fx* can have. We call *x* the "argument," and *fx* the "value for the argument *x*." When a function is what we call "continuous," the rough idea for which we are seeking a precise definition is that small differences in *x* shall correspond to small differences in *fx*, and if we make the differences in *x* small enough, we can make the differences in *fx* fall below any assigned amount. We do not want, if a function is to be continuous, that there shall be sudden jumps, so that, for some value of *x*, any change, however small, will make a change in *fx* which exceeds some assigned finite amount. The ordinary simple functions of mathematics have this property: it belongs, for example, to x^2, x^3, ... log *x*, sin *x*, and so on. But it is not at all difficult to define discontinuous functions. Take, as a non-mathematical example, "the place of birth of the youngest person living at time *t*." This is a function of *t*; its value is constant from the time of one person's birth to the time of the next birth, and then the value changes *suddenly* from one birthplace to the other. An analogous mathematical example would be "the integer next below *x*," where *x* is a real number. This function remains constant from one integer to the next, and then gives a sudden jump. The actual fact is that, though continuous functions are more familiar, they are the exceptions: there are infinitely more discontinuous functions than continuous ones.

Many functions are discontinuous for one or several values of the variable, but continuous for all other values. Take as an example sin $1/x$. The function sin θ passes through all values from -1 to 1 every time that θ passes from $-\pi/2$ to $\pi/2$, or from $\pi/2$ to $3\pi/2$, or generally from $(2n-1)\pi/2$ to $(2n+1)\pi/2$, where n is any integer. Now if we consider $1/x$ when x is very small, we see that as x diminishes $1/x$ grows faster and faster, so that it passes more and more quickly through the cycle of values from one multiple of $\pi/2$ to another as x becomes smaller and smaller. Consequently sin $1/x$ passes more and more quickly from -1 to 1 and back again, as x grows smaller. In fact, if we take any interval containing 0, say the interval from $-\varepsilon$ to $+\varepsilon$ where ε is some very small number, sin $1/x$ will go through an infinite number of oscillations in this interval, and we cannot diminish the oscillations by making the interval smaller. Thus round about the argument 0 the function is discontinuous. It is easy to manufacture functions which are discontinuous in several places, or in $\aleph_0$ places, or everywhere. Examples will be found in any book on the theory of functions of a real variable.

Proceeding now to seek a precise definition of what is meant by saying that a function is continuous for a given argument, when argument and value are both real numbers, let us first define a "neighbourhood" of a number x as all the numbers from $x-\varepsilon$ to $x+\varepsilon$, where ε is some number which, in important cases, will be very small. It is clear that continuity at a given point has to do with what happens in any neighbourhood of that point, however small.

What we desire is this: If a is the argument for which we wish our function to be continuous, let us first define a neighbourhood (α say) containing the value *fa* which the function has for the argument a; we desire that, if we take a sufficiently small neighbourhood containing a, all values for arguments throughout this neighbourhood shall be contained in the neighbourhood α, no matter how small we may have made α. That is to say, if we decree that our function is not to differ from *fa* by more than some very tiny amount, we can always find a stretch of real numbers, having a in the middle of it, such that throughout this stretch *fx* will not differ from *fa* by more than the prescribed tiny amount. And this is to remain true whatever tiny amount we may select. Hence we are led to the following definition:—

The function $f(x)$ is said to be "continuous" for the argument a if, for every positive number σ, different from 0, but as small as we please, there exists a positive number ε, different from 0, such that, for all values of δ which are numerically less [23] than ε, the difference $f(a+\delta)-f(a)$ is numerically less than σ.

[23] A number is said to be "numerically less" than ε when it lies between $-\varepsilon$ and $+\varepsilon$.

In this definition, σ first defines a neighbourhood of $f(a)$, namely, the neighbourhood from $f(a)-\sigma$ to $f(a)+\sigma$. The definition then proceeds to say that we can (by means of ε) define a neighbourhood, namely, that from $a-\varepsilon$ to $a+\varepsilon$, such that, for all arguments within this neighbourhood, the value of the function lies within the neighbourhood from $f(a)-\sigma$ to $f(a)+\sigma$. If this can be done, however σ may be chosen, the function is "continuous" for the argument a.

So far we have not defined the "limit" of a function for a given argument. If we had done so, we could have defined the continuity of a function differently: a function is continuous at a point where its value is the same as the limit of its values for approaches either from above or from below. But it is only the exceptionally "tame" function that has a definite limit as the argument approaches a given point. The general rule is that a function oscillates, and that, given any neighbourhood of a given argument, however small, a whole stretch of values will occur for arguments within this neighbourhood. As this is the general rule, let us consider it first.

Let us consider what may happen as the argument approaches some value a from below. That is to say, we wish to consider what happens for arguments contained in the interval from $a-\varepsilon$ to a, where ε is some number which, in important cases, will be very small.

The values of the function for arguments from $a-\varepsilon$ to a (a excluded) will be a set of real numbers which will define a certain section of the set of real numbers, namely, the section consisting of those numbers that are not greater than *all* the values for arguments from $a-\varepsilon$ to a. Given any number in this section, there are values at least as great as this number for arguments between $a-\varepsilon$ and a, *i.e.* for arguments that fall very little short of a (if ε is very small). Let us take all possible ε'*s* and all possible corresponding sections. The common part of all these sections we will call the "ultimate section" as the argument approaches a. To say that a number z belongs to the ultimate section is to say that, however small we may make ε, there are arguments between $a-\varepsilon$ and a for which the value of the function is not less than z.

We may apply exactly the same process to upper sections, *i.e.* to sections that go from some point up to the top, instead of from the bottom up to some point. Here we take those numbers that are not *less* than all the values for arguments from $a-\varepsilon$ to a; this defines an upper section which will vary as ε varies. Taking the common part of all such sections for all possible ε'*s*, we obtain the "ultimate upper section." To say that a number

z belongs to the ultimate upper section is to say that, however small we make ε, there are arguments between a−ε and a for which the value of the function is not *greater* than z.

If a term z belongs both to the ultimate section and to the ultimate upper section, we shall say that it belongs to the "ultimate oscillation." We may illustrate the matter by considering once more the function sin $1/x$ as x approaches the value 0. We shall assume, in order to fit in with the above definitions, that this value is approached from below.

Let us begin with the "ultimate section." Between −ε and 0, whatever ε may be, the function will assume the value 1 for certain arguments, but will never assume any greater value. Hence the ultimate section consists of all real numbers, positive and negative, up to and including 1; *i.e.* it consists of all negative numbers together with 0, together with the positive numbers up to and including 1.

Similarly the "ultimate upper section" consists of all positive numbers together with 0, together with the negative numbers down to and including −1.

Thus the "ultimate oscillation" consists of all real numbers from −1 to 1, both included.

We may say generally that the "ultimate oscillation" of a function as the argument approaches a from below consists of all those numbers x which are such that, however near we come to a, we shall still find values as great as x and values as small as x.

The ultimate oscillation may contain no terms, or one term, or many terms. In the first two cases the function has a definite limit for approaches from below. If the ultimate oscillation has one term, this is fairly obvious. It is equally true if it has none; for it is not difficult to prove that, if the ultimate oscillation is null, the boundary of the ultimate section is the same as that of the ultimate upper section, and may be defined as the limit of the function for approaches from below. But if the ultimate oscillation has many terms, there is no definite limit to the function for approaches from below. In this case we can take the lower and upper boundaries of the ultimate oscillation (*i.e.* the lower boundary of the ultimate upper section and the upper boundary of the ultimate section) as the lower and upper limits of its "ultimate" values for approaches from below. Similarly we obtain lower and upper limits of the "ultimate" values for approaches from above. Thus we have, in the general case, *four* limits to a function for approaches to a given argument. *The* limit for a given argument a only exists when all these four are equal, and is then their common value. If it is also the *value* for the argument a, the function is continuous for this argument. This may be taken as defining continuity: it is equivalent to our former definition.

We can define the limit of a function for a given argument (if it exists) without passing through the ultimate oscillation and the four limits of the general case. The definition proceeds, in that case, just as the earlier definition of continuity proceeded. Let us define the limit for approaches from below. If there is to be a definite limit for approaches to a from below, it is necessary and sufficient that, given any small number σ, two values for arguments sufficiently near to a (but both less than a) will differ by less than σ; *i.e.* if ε is sufficiently small, and our arguments both lie between a−ε and a (a excluded), then the difference between the values for these arguments will be less than σ. This is to hold for any σ, however small; in that case the function has a limit for approaches from below. Similarly we define the case when there is a limit for approaches from above. These two limits, even when both exist, need not be identical; and if they are identical, they still need not be identical with the *value* for the argument a. It is only in this last case that we call the function continuous for the argument a.

A function is called "continuous" (without qualification) when it is continuous for every argument.

Another slightly different method of reaching the definition of continuity is the following:—

Let us say that a function "ultimately converges into a class α" if there is some real number such that, for this argument and all arguments greater than this, the value of the function is a member of the class α. Similarly we shall say that a function "converges into α as the argument approaches *x* from below" if there is some argument *y* less than *x* such that throughout the interval from *y* (included) to *x* (excluded) the function has values which are members of α. We may now say that a function is continuous for the argument a, for which it has the value *fa*, if it satisfies four conditions, namely:—

(1) Given any real number less than *fa*, the function converges into the successors of this number as the argument approaches a from below;

(2) Given any real number greater than *fa*, the function converges into the predecessors of this number as the argument approaches a from below;

(3) and (4) Similar conditions for approaches to a from above.

The advantage of this form of definition is that it analyses the conditions of continuity into four, derived from considering arguments and values respectively greater or less than the argument and value for which continuity is to be defined.

We may now generalize our definitions so as to apply to series which are not numerical or known to be numerically measurable. The case of motion is a convenient one to bear in mind. There is a story by H. G. Wells which will illustrate, from the case of motion, the difference between the limit of a function for a given argument and its value for the same argument. The hero of the story, who possessed, without his knowledge, the power of realising his wishes, was being attacked by a policeman, but on ejaculating "Go to——" he found that the policeman disappeared. If *f*(*t*) was the policeman's position at time *t*, and t0 the moment of the ejaculation, the limit of the policeman's positions as *t* approached to t0 from below would be in contact with the hero, whereas the value for the argument t_0 was —. But such occurrences are supposed to be rare in the real world, and it is assumed, though without adequate evidence, that all motions are continuous, *i.e.* that, given any body, if *f*(*t*) is its position at time *t*, *f*(*t*) is a continuous function of *t*. It is the meaning of "continuity" involved in such statements which we now wish to define as simply as possible.

The definitions given for the case of functions where argument and value are real numbers can readily be adapted for more general use.

Let P and Q be two relations, which it is well to imagine serial, though it is not necessary to our definitions that they should be so. Let R be a one-many relation whose domain is contained in the field of P, while its converse domain is contained in the field of Q. Then R is (in a generalized sense) a function, whose arguments belong to the field of Q, while its values belong to the field of P. Suppose, for example, that we are dealing with a particle moving on a line: let Q be the time-series, P the series of points on our line from left to right, R the relation of the position of our particle on the line at time a to the time *a*, so that "the R of *a*" is its position at time *a*. This illustration may be borne in mind throughout our definitions.

We shall say that the function R is continuous for the argument *a* if, given any interval α on the P-series containing the value of the function for the argument *a*, there is an interval on the Q-series containing a not as an end-point and such that, throughout this interval, the function has values which are members of α. (We mean by an "interval" all the terms between any two; *i.e.* if *x* and *y* are two members of the field of P, and *x* has the relation P to *y*, we shall mean by the "P-interval *x* to *y*" all terms *z* such that *x* has the relation P to *z* and *z* has the relation P to *y*—together, when so stated, with *x* or *y* themselves.)

We can easily define the "ultimate section" and the "ultimate oscillation." To define the "ultimate section" for approaches to the argument a from below, take any argument *y* which precedes *a* (*i.e.* has the relation Q to *a*), take the values of the function for all arguments up to and including *y*, and form the section of P defined by these values, *i.e.* those members of the P-series which are earlier than or identical with some of these values. Form all such sections for all *y's* that precede a, and take their common part; this will be the ultimate section. The ultimate upper section and the ultimate oscillation are then defined exactly as in the previous case.

The adaptation of the definition of convergence and the resulting alternative definition of continuity offers no difficulty of any kind.

We say that a function R is "ultimately Q-convergent into α" if there is a member *y* of the converse domain of R and the field of Q such that the value of the function for the argument *y* and for any argument to which *y* has the relation Q is a member of α. We say that R "Q-converges into α as the argument approaches a given argument *a*" if there is a term *y* having the relation Q to *a* and belonging to the converse domain of R and such that the value of the function for any argument in the Q-interval from *y* (inclusive) to a (exclusive) belongs to α.

Of the four conditions that a function must fulfil in order to be continuous for the argument *a*, the first is, putting *b* for the value for the argument *a*:

Given any term having the relation P to *b*, R Q-converges into the successors of *b* (with respect to P) as the argument approaches a from below.

The second condition is obtained by replacing P by its converse; the third and fourth are obtained from the first and second by replacing Q by its converse.

There is thus nothing, in the notions of the limit of a function or the continuity of a function, that essentially involves number. Both can be defined generally, and many propositions about them can be proved for any two series (one being the argument-series and the other the value-series). It will be seen that the definitions do not involve infinitesimals. They involve infinite classes of intervals, growing smaller without any limit short of zero, but they do not involve any intervals that are not finite. This is analogous to the fact that if a line an inch long be halved, then halved again, and so on indefinitely, we never reach infinitesimals in this way: after *n* bisections, the length of our bit is $\frac{1}{2}^n$ of an inch; and this is finite whatever finite number *n* may be. The process of successive bisection does not lead to divisions whose ordinal number is infinite, since it is essentially a one-by-one process. Thus infinitesimals are not to be reached in this way. Confusions on such topics have had much to do with the difficulties which have been found in the discussion of infinity and continuity.

CHAPTER XII

SELECTIONS AND THE MULTIPLICATIVE AXIOM

In this chapter we have to consider an axiom which can be enunciated, but not proved, in terms of logic, and which is convenient, though not indispensable, in certain portions of mathematics. It is convenient, in the sense that many interesting propositions, which it seems natural to suppose true, cannot be proved without its help; but it is not indispensable, because even without those propositions the subjects in which they occur still exist, though in a somewhat mutilated form.

Before enunciating the multiplicative axiom, we must first explain the theory of selections, and the definition of multiplication when the number of factors may be infinite.

In defining the arithmetical operations, the only correct procedure is to construct an actual class (or relation, in the case of relation-numbers) having the required number of terms. This sometimes demands a certain amount of ingenuity, but it is essential in order to prove the existence of the number defined. Take, as the simplest example, the case of addition. Suppose we are given a cardinal number μ, and a class α which has μ terms. How shall we define $\mu+\mu$? For this purpose we must have *two* classes having μ terms, and they must not overlap. We can construct such classes from α in various ways, of which the following is perhaps the simplest: Form first all the ordered couples whose first term is a class consisting of a single member of α, and whose second term is the null-class; then, secondly, form all the ordered couples whose first term is the null-class and whose second term is a class consisting of a single member of α. These two classes of couples have no member in common, and the logical sum of the two classes will have $\mu+\mu$ terms. Exactly analogously we can define $\mu+\nu$, given that μ is the number of some class α and ν is the number of some class β.

Such definitions, as a rule, are merely a question of a suitable technical device. But in the case of multiplication, where the number of factors may be infinite, important problems arise out of the definition.

Multiplication when the number of factors is finite offers no difficulty. Given two classes α and β, of which the first has μ terms and the second ν terms, we can define $\mu\times\nu$ as the number of ordered couples that can be formed by choosing the first term out of α and the second out of β. It will be seen that this definition does not require that α and β should not overlap; it even remains adequate when α and β are identical. For example, let α be the class whose members are x_1, x_2, x_3. Then the class which is used to define the product $\mu\times\mu$ is the class of couples:

$$(x_1, x_1), (x_1, x_2), (x_1, x_3); (x_2, x_1), (x_2, x_2), (x_2, x_3); (x_3, x_1), (x_3, x_2), (x_3, x_3).$$

This definition remains applicable when μ or ν or both are infinite, and it can be extended step by step to three or four or any finite number of factors. No difficulty arises as regards this definition, except that it cannot be extended to an *infinite* number of factors.

The problem of multiplication when the number of factors may be infinite arises in this way: Suppose we have a class κ consisting of classes; suppose the number of terms in each of these classes is given. How shall we define the product of all these numbers? If we can frame our definition generally, it will be applicable whether κ is finite or infinite. It is to be observed that the problem is to be able to deal with the case when κ is infinite, not with the case when its members are. If κ is not infinite, the method defined above is just as applicable when its members are infinite as when they are finite. It is the case when κ is infinite, even though its members may be finite, that we have to find a way of dealing with.

The following method of defining multiplication generally is due to Dr Whitehead. It is explained and treated at length in *Principia Mathematica*, vol. *i*. *80ff., and vol. ii. *114.

Let us suppose to begin with that κ is a class of classes no two of which overlap—

say the constituencies in a country where there is no plural voting, each constituency being considered as a class of voters. Let us now set to work to choose one term out of each class to be its *representative*, as constituencies do when they elect members of Parliament, assuming that by law each constituency has to elect a man who is a voter in that constituency. We thus arrive at a class of representatives, who make up our Parliament, one being selected out of each constituency. How many different possible ways of choosing a Parliament are there? Each constituency can select any one of its voters, and therefore if there are μ voters in a constituency, it can make μ choices. The choices of the different constituencies are independent; thus it is obvious that, when the total number of constituencies is finite, the number of possible Parliaments is obtained by multiplying together the numbers of voters in the various constituencies. When we do not know whether the number of constituencies is finite or infinite, we may take the number of possible Parliaments as *defining* the product of the numbers of the separate constituencies. This is the method by which infinite products are defined. We must now drop our illustration, and proceed to exact statements.

Let κ be a class of classes, and let us assume to begin with that no two members of κ overlap, *i.e.* that if α and β are two different members of κ, then no member of the one is a member of the other. We shall call a class a "selection" from κ when it consists of just one term from each member of κ; *i.e.* μ is a "selection" from κ if every member of μ belongs to some member of κ, and if α be any member of κ, μ and α have exactly one term in common. The class of all "selections" from κ we shall call the "multiplicative class" of κ. The number of terms in the multiplicative class of κ, *i.e.* the number of possible selections from κ, is defined as the product of the numbers of the members of κ. This definition is equally applicable whether κ is finite or infinite.

Before we can be wholly satisfied with these definitions, we must remove the restriction that no two members of κ are to overlap. For this purpose, instead of defining first a class called a "selection," we will define first a relation which we will call a "selector." A relation R will be called a "selector" from κ if, from every member of κ, it picks out one term as the representative of that member, *i.e.* if, given any member α of κ, there is just one term *x* which is a member of α and has the relation R to α; and this is to be all that R does. The formal definition is:

A "selector" from a class of classes κ is a one-many relation, having κ for its converse domain, and such that, if *x* has the relation to α, then *x* is a member of α.

If R is a selector from κ, and α is a member of κ, and *x* is the term which has the relation R to α, we call *x* the "representative" of α in respect of the relation R.

A "selection" from κ will now be defined as the domain of a selector; and the multiplicative class, as before, will be the class of selections.

But when the members of κ overlap, there may be more selectors than selections, since a term *x* which belongs to two classes α and β may be selected once to represent α and once to represent β, giving rise to different selectors in the two cases, but to the same selection. For purposes of defining multiplication, it is the selectors we require rather than the selections. Thus we define:

"The product of the numbers of the members of a class of classes κ" is the number of selectors from κ.

We can define exponentiation by an adaptation of the above plan. We might, of course, define μν as the number of selectors from ν classes, each of which has μ terms. But there are objections to this definition, derived from the fact that the multiplicative axiom (of which we shall speak shortly) is unnecessarily involved if it is adopted. We adopt instead the following construction:—

Let α be a class having μ terms, and β a class having ν terms. Let y be a member of β, and form the class of all ordered couples that have y for their second term and a member of α for their first term. There will be μ such couples for a given y, since any member of α may be chosen for the first term, and α has μ members. If we now form all the classes of this sort that result from varying y, we obtain altogether ν classes, since y may be any member of β, and β has ν members. These ν classes are each of them a class of couples, namely, all the couples that can be formed of a variable member of α and a fixed member of β. We define μ^ν as the number of selectors from the class consisting of these ν classes. Or we may equally well define μ^ν as the number of selections, for, since our classes of couples are mutually exclusive, the number of selectors is the same as the number of selections. A selection from our class of classes will be a set of ordered couples, of which there will be exactly one having any given member of β for its second term, and the first term may be any member of α. Thus μ^ν is defined by the selectors from a certain set of ν classes each having μ terms, but the set is one having a certain structure and a more manageable composition than is the case in general. The relevance of this to the multiplicative axiom will appear shortly.

What applies to exponentiation applies also to the product of two cardinals. We might define "$\mu\times\nu$" as the sum of the numbers of ν classes each having μ terms, but we prefer to define it as the number of ordered couples to be formed consisting of a member of α followed by a member of β, where α has μ terms and β has ν terms. This definition, also, is designed to evade the necessity of assuming the multiplicative axiom.

With our definitions, we can prove the usual formal laws of multiplication and exponentiation. But there is one thing we cannot prove: we cannot prove that a product is only zero when one of its factors is zero. We can prove this when the number of factors is finite, but not when it is infinite. In other words, we cannot prove that, given a class of classes none of which is null, there must be selectors from them; or that, given a class of mutually exclusive classes, there must be at least one class consisting of one term out of each of the given classes. These things cannot be proved; and although, at first sight, they seem obviously true, yet reflection brings gradually increasing doubt, until at last we become content to register the assumption and its consequences, as we register the axiom of parallels, without assuming that we can know whether it is true or false. The assumption, loosely worded, is that selectors and selections exist when we should expect them. There are many equivalent ways of stating it precisely. We may begin with the following:—

"Given any class of mutually exclusive classes, of which none is null, there is at least one class which has exactly one term in common with each of the given classes."

This proposition we will call the "multiplicative axiom." [24] We will first give various equivalent forms of the proposition, and then consider certain ways in which its truth or falsehood is of interest to mathematics.

[24] See *Principia Mathematica*, vol. *i*. *88. Also vol. iii. *257–258.

The multiplicative axiom is equivalent to the proposition that a product is only zero when at least one of its factors is zero; *i.e.* that, if any number of cardinal numbers be multiplied together, the result cannot be 0 unless one of the numbers concerned is 0.

The multiplicative axiom is equivalent to the proposition that, if R be any relation, and κ any class contained in the converse domain of R, then there is at least one one-many relation implying R and having κ for its converse domain.

The multiplicative axiom is equivalent to the assumption that if α be any class, and κ all the sub-classes of α with the exception of the null-class, then there is at least one selector from κ. This is the form in which the axiom was first brought to the notice of the learned world by Zermelo, in his "Beweis, dass jede Menge wohlgeordnet werden kann." [25] Zermelo regards the axiom as an unquestionable truth. It must be confessed that, until he made it explicit, mathematicians had used it without a qualm; but it would seem that they had done so unconsciously. And the credit due to Zermelo for having made it explicit is entirely independent of the question whether it is true or false.

[25] *Mathematische Annalen*, vol. lix. pp. 514–6. In this form we shall speak of it as Zermelo'*s* axiom.

The multiplicative axiom has been shown by Zermelo, in the above-mentioned proof, to be equivalent to the proposition that every class can be well-ordered, *i.e.* can be arranged in a series in which every sub-class has a first term (except, of course, the null-class). The full proof of this proposition is difficult, but it is not difficult to see the general principle upon which it proceeds. It uses the form which we call "Zermelo'*s* axiom," *i.e.* it assumes that, given any class α, there is at least one one-many relation R whose converse domain consists of all existent sub-classes of α and which is such that, if x has the relation R to ξ, then x is a member of ξ. Such a relation picks out a "representative" from each sub-class; of course, it will often happen that two sub-classes have the same representative. What Zermelo does, in effect, is to count off the members of α, one by one, by means of R and transfinite induction. We put first the representative of α; call it x_1. Then take the representative of the class consisting of all of α except x_1; call it x_2. It must be different from x_1, because every representative is a member of its class, and x_1 is shut out from this class. Proceed similarly to take away x_2, and let x_3 be the representative of what is left. In this way we first obtain a progression x_1, x_2, ... xn, ..., assuming that α is not finite. We then take away the whole progression; let x_ω be the representative of what is left of α. In this way we can go on until nothing is left. The successive representatives will form a well-ordered series containing all the members of α. (The above is, of course, only a hint of the general lines of the proof.) This proposition is called "Zermelo'*s* theorem."

The multiplicative axiom is also equivalent to the assumption that of any two cardinals which are not equal, one must be the greater. If the axiom is false, there will be cardinals μ and ν such that μ is neither less than, equal to, nor greater than ν. We have seen that $\aleph_1$ and $2^{\aleph_0}$ possibly form an instance of such a pair.

Many other forms of the axiom might be given, but the above are the most important of the forms known at present. As to the truth or falsehood of the axiom in any of its forms, nothing is known at present.

The propositions that depend upon the axiom, without being known to be equivalent to it, are numerous and important. Take first the connection of addition and multiplication. We naturally think that the sum of ν mutually exclusive classes, each having μ terms, must have μ×ν terms. When ν is finite, this can be proved. But when ν is infinite, it cannot be proved without the multiplicative axiom, except where, owing to some special circumstance, the existence of certain selectors can be proved. The way the multiplicative axiom enters in is as follows: Suppose we have two sets of ν mutually exclusive classes, each having μ terms, and we wish to prove that the sum of one set has as many terms as the sum of the other. In order to prove this, we must establish a one-one relation. Now, since there are in each case ν classes, there is some one-one relation between the two sets of classes; but what we want is a one-one relation between their terms. Let us consider some one-one relation S between the classes. Then if κ and λ are the two sets of classes, and α is some member of κ, there will be a member β of λ which will be the correlate of α with respect to S. Now α and β each have μ terms, and are therefore similar. There are, accordingly, one-one correlations of α and β. The trouble is that there are so many. In order to obtain a one-one correlation of the sum of κ with the sum of λ, we have to pick out *one selection* from a set of classes of correlators, one class of the set being all the one-one correlators of α with β. If κ and λ are infinite, we cannot in general know that such a selection exists, unless we can know that the multiplicative axiom is true. Hence we cannot establish the usual kind of connection between addition and multiplication.

This fact has various curious consequences. To begin with, we know that $\aleph_0^2=\aleph_0\times\aleph_0=\aleph_0$. It is commonly inferred from this that the sum of $\aleph_0$ classes each having $\aleph_0$ members must itself have $\aleph0$ members, but this inference is fallacious, since we do not know that the number of terms in such a sum is $\aleph_0\times\aleph_0$, nor consequently that it is $\aleph_0$. This has a bearing upon the theory of transfinite ordinals. It is easy to prove that an ordinal which has $\aleph_0$ predecessors must be one of what Cantor calls the "second class," *i.e.* such that a series having this ordinal number will have $\aleph_0$ terms in its field. It is also easy to see that, if we take any progression of ordinals of the second class, the predecessors of their limit form at most the sum of $\aleph_0$ classes each having $\aleph_0$ terms. It is inferred thence—fallaciously, unless the multiplicative axiom is true—that the predecessors of the limit are $\aleph_0$ in number, and therefore that the limit is a number of the "second class." That is to say, it is supposed to be proved that any progression of ordinals of the second class has a limit which is again an ordinal of the second class. This proposition, with the corollary that ω_1 (the smallest ordinal of the third class) is not the limit of any progression, is involved in most of the recognised theory of ordinals of the second class. In view of the way in which the multiplicative axiom is involved, the proposition and its corollary cannot be regarded as proved. They may be true, or they may not. All that can be said at present is that we do not know. Thus the greater part of the theory of ordinals of the second class must be regarded as unproved.

Another illustration may help to make the point clearer. We know that $2\times\aleph_0=\aleph_0$. Hence we might suppose that the sum of $\aleph_0$ pairs must have $\aleph_0$ terms. But this, though we can prove that it is sometimes the case, cannot be proved to happen always unless we assume the multiplicative axiom. This is illustrated by the millionaire who bought a pair of socks whenever he bought a pair of boots, and never at any other time, and who had such a passion for buying both that at last he had $\aleph_0$ pairs of boots and $\aleph_0$ pairs of socks. The problem is: How many boots had he, and how many socks? One would naturally suppose that he had twice as many boots and twice as many socks as he had pairs of each, and that therefore he had $\aleph_0$ of each, since that number is not increased by doubling. But this is an instance of the difficulty, already noted, of connecting the sum of ν classes each having μ terms with $\mu\times\nu$. Sometimes this can be done, sometimes it cannot. In our case it can be done with the boots, but not with the socks, except by some very artificial device. The reason for the difference is this: Among boots we can distinguish right and left, and therefore we can make a selection of one out of each pair, namely, we can choose all the right boots or all the left boots; but with socks no such principle of selection suggests itself, and we cannot be sure, unless we assume the multiplicative axiom, that there is any class consisting of one sock out of each pair. Hence the problem.

We may put the matter in another way. To prove that a class has $\aleph_0$ terms, it is necessary and sufficient to find some way of arranging its terms in a progression. There is no difficulty in doing this with the boots. The pairs are given as forming an $\aleph_0$, and therefore as the field of a progression. Within each pair, take the left boot first and the right second, keeping the order of the pair unchanged; in this way we obtain a progression of all the boots. But with the socks we shall have to choose arbitrarily, with each pair, which to put first; and an infinite number of arbitrary choices is an impossibility. Unless we can find a *rule* for selecting, *i.e.* a relation which is a selector, we do not know that a selection is even theoretically possible. Of course, in the case of objects in space, like socks, we always can find some principle of selection. For example, take the centres of mass of the socks: there will be points p in space such that, with any pair, the centres of mass of the two socks are not both at exactly the same distance from p; thus we can choose, from each pair, that sock which has its centre of mass nearer to p. But there is no theoretical reason why a method of selection such as this should always be possible, and the case of the socks, with a little goodwill on the part of the reader, may serve to show how a selection might be impossible.

It is to be observed that, if it *were* impossible to select one out of each pair of socks, it would follow that the socks *could* not be arranged in a progression, and therefore that there were not $\aleph_0$ of them. This case illustrates that, if μ is an infinite number, one set of μ pairs may not contain the same number of terms as another set of μ pairs; for, given $\aleph_0$ pairs of boots, there are certainly $\aleph_0$ boots, but we cannot be sure of this in the case of the socks unless we assume the multiplicative axiom or fall back upon some fortuitous geometrical method of selection such as the above.

Another important problem involving the multiplicative axiom is the relation of reflexiveness to non-inductiveness. It will be remembered that in Chapter VIII. we pointed out that a reflexive number must be non-inductive, but that the converse (so far as is known at present) can only be proved if we assume the multiplicative axiom. The way in which this comes about is as follows:—

It is easy to prove that a reflexive class is one which contains sub-classes having $\aleph_0$ terms. (The class may, of course, itself have $\aleph_0$ terms.) Thus we have to prove, if we can, that, given any non-inductive class, it is possible to choose a progression out of its terms. Now there is no difficulty in showing that a non-inductive class must contain more terms than any inductive class, or, what comes to the same thing, that if α is a non-inductive class and ν is any inductive number, there are sub-classes of α that have ν terms. Thus we can form sets of finite sub-classes of α: First one class having no terms, then classes having 1 term (as many as there are members of α), then classes having 2 terms, and so on. We thus get a progression of sets of sub-classes, each set consisting of all those that have a certain given finite number of terms. So far we have not used the multiplicative axiom, but we have only proved that the number of collections of sub-classes of α is a reflexive number, *i.e.* that, if μ is the number of members of α, so that 2^μ is the number of sub-classes of α and 2^{2^μ} is the number of collections of sub-classes, then, provided μ is not inductive, 2^{2^μ} must be reflexive. But this is a long way from what we set out to prove.

In order to advance beyond this point, we must employ the multiplicative axiom. From each set of sub-classes let us choose out one, omitting the sub-class consisting of the null-class alone. That is to say, we select one sub-class containing one term, α_1, say; one containing two terms, α_2, say; one containing three, α_3, say; and so on. (We can do this if the multiplicative axiom is assumed; otherwise, we do not know whether we can always do it or not.) We have now a progression α_1, α_2, α_3, ... of sub-classes of α, instead of a progression of collections of sub-classes; thus we are one step nearer to our goal. We now know that, assuming the multiplicative axiom, if μ is a non-inductive number, 2^μ must be a reflexive number.

The next step is to notice that, although we cannot be sure that new members of α come in at any one specified stage in the progression α_1, α_2, α_3, ... we can be sure that new members keep on coming in from time to time. Let us illustrate. The class α_1, which consists of one term, is a new beginning; let the one term be x_1. The class α_2, consisting of two terms, may or may not contain x_1; if it does, it introduces one new term; and if it does not, it must introduce two new terms, say x_2, x_3. In this case it is possible that α_3 consists of x_1, x_2, x_3, and so introduces no new terms, but in that case α_4 must introduce a new term. The first ν classes α_1, α_2, α_3, ... α_ν contain, at the very most, $1+2+3+ \dots +\nu$ terms, *i.e.* $\nu(\nu+1)/2$ terms; thus it would be possible, if there were no repetitions in the first ν classes, to go on with only repetitions from the $(\nu+1)^{th}$ class to the $\nu(\nu+1)/2^{\text{th}}$ class. But by that time the old terms would no longer be sufficiently numerous to form a next class with the right number of members, *i.e.* $\nu(\nu+1)/2+1$, therefore new terms must come in at this point if not sooner. It follows that, if we omit from our progression α_1, α_2, α_3, ... all those classes that are composed entirely of members that have occurred in previous classes, we shall still have a progression. Let our new progression be called β_1, β_2, β_3 ... (We shall have $\alpha_1=\beta_1$ and $\alpha_2=\beta_2$, because α_1 and α_2 must introduce new terms. We may or may not have $\alpha_3=\beta_3$, but, speaking generally, β_μ will be $\alpha\nu$, where ν is some number greater than μ; *i.e.* the β'*s* are some of the α'*s*.) Now these β'*s* are such that any one of them, say β_μ, contains members which have not occurred in any of the previous β'*s*. Let γ_μ be the part of β_μ which consists of new members. Thus we get a new progression γ_1, γ_2, γ_3, ... (Again γ_1 will be identical with β_1 and with α_1; if α_2 does not contain the one member of α_1, we shall have $\gamma_2=\beta_2=\alpha_2$, but if α_2 does contain this one member, γ_2 will consist of the other member of α_2.) This new progression of γ'*s* consists of mutually exclusive classes. Hence a selection from them will be a progression; *i.e.* if x_1 is the member of γ_1, x_2 is a member of γ_2, x_3 is a member of γ_3, and so on; then x_1, x_2, x_3, ... is a progression, and is a sub-class of α. Assuming the multiplicative axiom, such a selection can be made. Thus by twice using this axiom we can prove that, if the axiom is true, every non-inductive cardinal must be reflexive. This could also be deduced from Zermelo'*s* theorem, that, if the axiom is true, every class can be well-ordered; for a well-ordered series must have either a finite or a reflexive number of terms in its field.

There is one advantage in the above direct argument, as against deduction from Zermelo'*s* theorem, that the above argument does not demand the universal truth of the multiplicative axiom, but only its truth as applied to a set of $\aleph_0$ classes. It may happen that the axiom holds for $\aleph_0$ classes, though not for larger numbers of classes. For this reason it is better, when it is possible, to content ourselves with the more restricted assumption. The assumption made in the above direct argument is that a product of $\aleph_0$ factors is never zero unless one of the factors is zero. We may state this assumption in the form: "$\aleph_0$ is a *multipliable* number," where a number ν is defined as "multipliable" when a product of ν factors is never zero unless one of the factors is zero. We can *prove* that a *finite* number is always multipliable, but we cannot prove that any infinite number is so. The multiplicative axiom is equivalent to the assumption that *all* cardinal numbers are multipliable. But in order to identify the reflexive with the non-inductive, or to deal with the problem of the boots and socks, or to show that any progression of numbers of the second class is of the second class, we only need the very much smaller assumption that

$\aleph_0$ is multipliable.

It is not improbable that there is much to be discovered in regard to the topics discussed in the present chapter. Cases may be found where propositions which seem to involve the multiplicative axiom can be proved without it. It is conceivable that the multiplicative axiom in its general form may be shown to be false. From this point of view, Zermelo's theorem offers the best hope: the continuum or some still more dense series *might* be proved to be incapable of having its terms well-ordered, which would prove the multiplicative axiom false, in virtue of Zermelo's theorem. But so far, no method of obtaining such results has been discovered, and the subject remains wrapped in obscurity.

CHAPTER XIII

THE AXIOM OF INFINITY AND LOGICAL TYPES

The axiom of infinity is an assumption which may be enunciated as follows:—

"If n be any inductive cardinal number, there is at least one class of individuals having n terms."

If this is true, it follows, of course, that there are many classes of individuals having n terms, and that the total number of individuals in the world is not an inductive number. For, by the axiom, there is at least one class having $n+1$ terms, from which it follows that there are many classes of n terms and that n is not the number of individuals in the world. Since n is *any* inductive number, it follows that the number of individuals in the world must (if our axiom be true) exceed any inductive number. In view of what we found in the preceding chapter, about the possibility of cardinals which are neither inductive nor reflexive, we cannot infer from our axiom that there are at least $\aleph_0$ individuals, unless we assume the multiplicative axiom. But we do know that there are at least $\aleph_0$ classes of classes, since the inductive cardinals are classes of classes, and form a progression if our axiom is true.

The way in which the need for this axiom arises may be explained as follows. One of Peano's assumptions is that no two inductive cardinals have the same successor, *i.e.* that we shall not have $m+1=n+1$ unless $m=n$, if m and n are inductive cardinals. In Chapter VIII. we had occasion to use what is virtually the same as the above assumption of Peano's, namely, that, if n is an inductive cardinal, n is not equal to $n+1$. It might be thought that this could be proved. We can prove that, if α is an inductive class, and n is the number of members of α, then n is not equal to $n+1$. This proposition is easily proved by induction, and might be thought to imply the other. But in fact it does not, since there might be no such class as α. What it does imply is this: If n is an inductive cardinal such that there is at least one class having n members, then n is not equal to $n+1$. The axiom of infinity assures us (whether truly or falsely) that there are classes having n members, and thus enables us to assert that n is not equal to $n+1$. But without this axiom we should be left with the possibility that n and $n+1$ might both be the null-class.

Let us illustrate this possibility by an example: Suppose there were exactly nine individuals in the world. (As to what is meant by the word "individual," I must ask the reader to be patient.) Then the inductive cardinals from 0 up to 9 would be such as we expect, but 10 (defined as 9+1) would be the null-class. It will be remembered that $n+1$ may be defined as follows: $n+1$ is the collection of all those classes which have a term x such that, when x is taken away, there remains a class of n terms. Now applying this definition, we see that, in the case supposed, 9+1 is a class consisting of no classes, *i.e.* it is the null-class. The same will be true of 9+2, or generally of 9+n, unless n is zero. Thus 10 and all subsequent inductive cardinals will all be identical, since they will all be the null-class. In such a case the inductive cardinals will not form a progression, nor will it be true that no two have the same successor, for 9 and 10 will both be succeeded by the null-class (10 being itself the null-class). It is in order to prevent such arithmetical catastrophes that we require the axiom of infinity.

As a matter of fact, so long as we are content with the arithmetic of finite integers, and do not introduce either infinite integers or infinite classes or series of finite integers or ratios, it is possible to obtain all desired results without the axiom of infinity. That is to say, we can deal with the addition, multiplication, and exponentiation of finite integers and of ratios, but we cannot deal with infinite integers or with irrationals. Thus the theory of the transfinite and the theory of real numbers fails us. How these various results come about must now be explained.

Assuming that the number of individuals in the world is n, the number of classes of individuals will be 2^n. This is in virtue of the general proposition mentioned in Chapter VIII. that the number of classes contained in a class which has n members is 2^n. Now 2^n is always greater than n. Hence the number of classes in the world is greater than the number of individuals. If, now, we suppose the number of individuals to be 9, as we did just now, the number of classes will be 2^9, *i.e.* 512. Thus if we take our numbers as being applied to the counting of classes instead of to the counting of individuals, our arithmetic will be normal until we reach 512: the first number to be null will be 513. And if we advance to classes of classes we shall do still better: the number of them will be 2^{512}, a number which is so large as to stagger imagination, since it has about 153 digits. And if we advance to classes of classes of classes, we shall obtain a number represented by 2 raised to a power which has about 153 digits; the number of digits in this number will be about three times 10^{152}. In a time of paper shortage it is undesirable to write out this number, and if we want larger ones we can obtain them by travelling further along the logical hierarchy. In this way any assigned inductive cardinal can be made to find its place among numbers which are not null, merely by travelling along the hierarchy for a sufficient distance. [26]

[26] On this subject see *Principia Mathematica*, vol. ii. *120ff. On the corresponding problems as regards ratio, see *ibid.*, vol. iii. *303ff.

As regards ratios, we have a very similar state of affairs. If a ratio μ/ν is to have the expected properties, there must be enough objects of whatever sort is being counted to insure that the null-class does not suddenly obtrude itself. But this can be insured, for any given ratio μ/ν, without the axiom of infinity, by merely travelling up the hierarchy a sufficient distance. If we cannot succeed by counting individuals, we can try counting classes of individuals; if we still do not succeed, we can try classes of classes, and so on. Ultimately, however few individuals there may be in the world, we shall reach a stage where there are many more than μ objects, whatever inductive number μ may be. Even if there were no individuals at all, this would still be true, for there would then be one class, namely, the null-class, 2 classes of classes (namely, the null-class of classes and the class whose only member is the null-class of individuals), 4 classes of classes of classes, 16 at the next stage, 65,536 at the next stage, and so on. Thus no such assumption as the axiom of infinity is required in order to reach any given ratio or any given inductive cardinal.

It is when we wish to deal with the whole class or series of inductive cardinals or of ratios that the axiom is required. We need the whole class of inductive cardinals in order to establish the existence of $\aleph_0$, and the whole series in order to establish the existence of progressions: for these results, it is necessary that we should be able to make a single class or series in which no inductive cardinal is null. We need the whole series of ratios in order of magnitude in order to define real numbers as segments: this definition will not give the desired result unless the series of ratios is compact, which it cannot be if the total number of ratios, at the stage concerned, is finite.

It would be natural to suppose—as I supposed myself in former days—that, by means of constructions such as we have been considering, the axiom of infinity could be *proved*. It may be said: Let us assume that the number of individuals is n, where n may be 0 without spoiling our argument; then if we form the complete set of individuals, classes, classes of classes, etc., all taken together, the number of terms in our whole set will be

$$n+2^n+2^{2^n} \ldots \textit{ad inf.},$$

which is $\aleph_0$. Thus taking all kinds of objects together, and not confining ourselves to objects of any one type, we shall certainly obtain an infinite class, and shall therefore not need the axiom of infinity. So it might be said.

Now, before going into this argument, the first thing to observe is that there is an air of hocus-pocus about it: something reminds one of the conjurer who brings things out of the hat. The man who has lent his hat is quite sure there wasn't a live rabbit in it before, but he is at a loss to say how the rabbit got there. So the reader, if he has a robust sense of reality, will feel convinced that it is impossible to manufacture an infinite collection out of a finite collection of individuals, though he may be unable to say where the flaw is in the above construction. It would be a mistake to lay too much stress on such feelings of hocus-pocus; like other emotions, they may easily lead us astray. But they afford a *prima facie* ground for scrutinizing very closely any argument which arouses them. And when the above argument is scrutinized it will, in my opinion, be found to be fallacious, though the fallacy is a subtle one and by no means easy to avoid consistently.

The fallacy involved is the fallacy which may be called "confusion of types." To explain the subject of "types" fully would require a whole volume; moreover, it is the purpose of this book to avoid those parts of the subjects which are still obscure and controversial, isolating, for the convenience of beginners, those parts which can be accepted as embodying mathematically ascertained truths. Now the theory of types emphatically does not belong to the finished and certain part of our subject: much of this theory is still inchoate, confused, and obscure. But the need of *some* doctrine of types is less doubtful than the precise form the doctrine should take; and in connection with the axiom of infinity it is particularly easy to see the necessity of some such doctrine.

This necessity results, for example, from the "contradiction of the greatest cardinal." We saw in Chapter VIII. that the number of classes contained in a given class is always greater than the number of members of the class, and we inferred that there is no greatest cardinal number. But if we could, as we suggested a moment ago, add together into one class the individuals, classes of individuals, classes of classes of individuals, etc., we should obtain a class of which its own sub-classes would be members. The class consisting of all objects that can be counted, of whatever sort, must, if there be such a class, have a cardinal number which is the greatest possible. Since all its sub-classes will be members of it, there cannot be more of them than there are members. Hence we arrive at a contradiction.

When I first came upon this contradiction, in the year 1901, I attempted to discover some flaw in Cantor's proof that there is no greatest cardinal, which we gave in Chapter VIII. Applying this proof to the supposed class of all imaginable objects, I was led to a new and simpler contradiction, namely, the following:—

The comprehensive class we are considering, which is to embrace everything, must embrace itself as one of its members. In other words, if there is such a thing as "everything," then "everything" is something, and is a member of the class "everything." But normally a class is not a member of itself. Mankind, for example, is not a man. Form now the assemblage of all classes which are not members of themselves. This is a class: is it a member of itself or not? If it is, it is one of those classes that are not members of themselves, *i.e.* it is not a member of itself. If it is not, it is not one of those classes that are not members of themselves, *i.e.* it is a member of itself. Thus of the two hypotheses—that it is, and that it is not, a member of itself—each implies its contradictory. This is a contradiction.

There is no difficulty in manufacturing similar contradictions ad lib. The solution of such contradictions by the theory of types is set forth fully in *Principia Mathematica*, [27] and also, more briefly, in articles by the present author in the *American Journal of Mathematics* [28] and in the *Revue de Metaphysique et de Morale*. [29] For the present an outline of the solution must suffice.

[27] Vol. *i.*, Introduction, chap. ii., *12 and *20; vol. ii., Prefatory Statement.
[28] "Mathematical Logic as based on the Theory of Types," vol. xxx., 1908, pp. 222–262.
[29] "Les paradoxes de la logique," 1906, pp. 627–650.

The fallacy consists in the formation of what we may call "impure" classes, *i.e.* classes which are not pure as to "type." As we shall see in a later chapter, classes are logical fictions, and a statement which appears to be about a class will only be significant if it is capable of translation into a form in which no mention is made of the class. This places a limitation upon the ways in which what are nominally, though not really, names for classes can occur significantly: a sentence or set of symbols in which such pseudo-names occur in wrong ways is not false, but strictly devoid of meaning. The supposition that a class is, or that it is not, a member of itself is meaningless in just this way. And more generally, to suppose that one class of individuals is a member, or is not a member, of another class of individuals will be to suppose nonsense; and to construct symbolically any class whose members are not all of the same grade in the logical hierarchy is to use symbols in a way which makes them no longer symbolize anything.

Thus if there are n individuals in the world, and 2^n classes of individuals, we cannot form a new class, consisting of both individuals and classes and having $n+2^n$ members. In this way the attempt to escape from the need for the axiom of infinity breaks down. I do not pretend to have explained the doctrine of types, or done more than indicate, in rough outline, why there is need of such a doctrine. I have aimed only at saying just so much as was required in order to show that we cannot *prove* the existence of infinite numbers and classes by such conjurer's methods as we have been examining. There remain, however, certain other possible methods which must be considered.

Various arguments professing to prove the existence of infinite classes are given in the *Principles of Mathematics*, §339 (p. 357). In so far as these arguments assume that, if

n is an inductive cardinal, n is not equal to $n+1$, they have been already dealt with. There is an argument, suggested by a passage in Plato's *Parmenides*, to the effect that, if there is such a number as 1, then 1 has being; but 1 is not identical with being, and therefore 1 and being are two, and therefore there is such a number as 2, and 2 together with 1 and being gives a class of three terms, and so on. This argument is fallacious, partly because "being" is not a term having any definite meaning, and still more because, if a definite meaning were invented for it, it would be found that numbers do not have being—they are, in fact, what are called "logical fictions," as we shall see when we come to consider the definition of classes.

The argument that the number of numbers from 0 to n (both inclusive) is $n+1$ depends upon the assumption that up to and including n no number is equal to its successor, which, as we have seen, will not be always true if the axiom of infinity is false. It must be understood that the equation $n=n+1$, which might be true for a finite n if n exceeded the total number of individuals in the world, is quite different from the same equation as applied to a reflexive number. As applied to a reflexive number, it means that, given a class of n terms, this class is "similar" to that obtained by adding another term. But as applied to a number which is too great for the actual world, it merely means that there is no class of n individuals, and no class of $n+1$ individuals; it does not mean that, if we mount the hierarchy of types sufficiently far to secure the existence of a class of n terms, we shall then find this class "similar" to one of $n+1$ terms, for if n is inductive this will not be the case, quite independently of the truth or falsehood of the axiom of infinity.

There is an argument employed by both Bolzano [30] and Dedekind [31] to prove the existence of reflexive classes. The argument, in brief, is this: An object is not identical with the idea of the object, but there is (at least in the realm of being) an idea of any object. The relation of an object to the idea of it is one-one, and ideas are only some among objects. Hence the relation "idea of" constitutes a reflexion of the whole class of objects into a part of itself, namely, into that part which consists of ideas. Accordingly, the class of objects and the class of ideas are both infinite. This argument is interesting, not only on its own account, but because the mistakes in it (or what I judge to be mistakes) are of a kind which it is instructive to note. The main error consists in assuming that there is an idea of every object. It is, of course, exceedingly difficult to decide what is meant by an "idea"; but let us assume that we know. We are then to suppose that, starting (say) with Socrates, there is the idea of Socrates, and then the idea of the idea of Socrates, and so on *ad inf.* Now it is plain that this is not the case in the sense that all these ideas have actual empirical existence in people's minds. Beyond the third or fourth stage they become mythical. If the argument is to be upheld, the "ideas" intended must be Platonic ideas laid up in heaven, for certainly they are not on earth. But then it at once becomes doubtful whether there are such ideas. If we are to know that there are, it must be on the basis of some logical theory, proving that it is necessary to a thing that there should be an idea of it. We certainly cannot obtain this result empirically, or apply it, as Dedekind does, to "meine Gedankenwelt"—the world of my thoughts.

[30] Bolzano, *Paradoxien des Unendlichen*, 13.
[31] Dedekind, *Was sind und was sollen die Zahlen?* No. 66.

If we were concerned to examine fully the relation of idea and object, we should have to enter upon a number of psychological and logical inquiries, which are not relevant to our main purpose. But a few further points should be noted. If "idea" is to be understood logically, it may be *identical* with the object, or it may stand for a *description* (in the sense to be explained in a subsequent chapter). In the former case the argument fails, because it was essential to the proof of reflexiveness that object and idea should be distinct. In the second case the argument also fails, because the relation of object and description is not one-one: there are innumerable correct descriptions of any given object. Socrates (*e.g.*) may be described as "the master of Plato," or as "the philosopher who drank the hemlock," or as "the husband of Xantippe." If—to take up the remaining hypothesis—"idea" is to be interpreted psychologically, it must be maintained that there is not any one definite psychological entity which could be called *the* idea of the object: there are innumerable beliefs and attitudes, each of which could be called *an* idea of the object in the sense in which we might say "my idea of Socrates is quite different from yours," but there is not any central entity (except Socrates himself) to bind together various "ideas of Socrates," and thus there is not any such one-one relation of idea and object as the argument supposes. Nor, of course, as we have already noted, is it true psychologically that there are ideas (in however extended a sense) of more than a tiny proportion of the things in the world. For all these reasons, the above argument in favour of the logical existence of reflexive classes must be rejected.

It might be thought that, whatever may be said of *logical* arguments, the *empirical* arguments derivable from space and time, the diversity of colours, etc., are quite sufficient to prove the actual existence of an infinite number of particulars. I do not believe this. We have no reason except prejudice for believing in the infinite extent of space and time, at any rate in the sense in which space and time are physical facts, not mathematical fictions. We naturally regard space and time as continuous, or, at least, as compact; but this again is mainly prejudice. The theory of "quanta" in physics, whether true or false, illustrates the fact that physics can never afford proof of continuity, though it might quite possibly afford disproof. The senses are not sufficiently exact to distinguish between continuous motion and rapid discrete succession, as anyone may discover in a cinema. A world in which all motion consisted of a series of small finite jerks would be empirically indistinguishable from one in which motion was continuous. It would take up too much space to defend these theses adequately; for the present I am merely suggesting them for the reader's consideration. If they are valid, it follows that there is no empirical reason for believing the number of particulars in the world to be infinite, and that there never can be; also that there is at present no empirical reason to believe the number to be finite, though it is theoretically conceivable that someday there might be evidence pointing, though not conclusively, in that direction.

From the fact that the infinite is not self-contradictory, but is also not demonstrable logically, we must conclude that nothing can be known *a priori* as to whether the number of things in the world is finite or infinite. The conclusion is, therefore, to adopt a Leibnizian phraseology, that some of the possible worlds are finite, some infinite, and we have no means of knowing to which of these two kinds our actual world belongs. The axiom of infinity will be true in some possible worlds and false in others; whether it is true or false in this world, we cannot tell.

Throughout this chapter the synonyms "individual" and "particular" have been used without explanation. It would be impossible to explain them adequately without a longer disquisition on the theory of types than would be appropriate to the present work, but a few words before we leave this topic may do something to diminish the obscurity which would otherwise envelop the meaning of these words.

In an ordinary statement we can distinguish a verb, expressing an attribute or relation, from the substantives which express the subject of the attribute or the terms of the relation. "Cæsar lived" ascribes an attribute to Cæsar; "Brutus killed Cæsar" expresses a relation between Brutus and Cæsar. Using the word "subject" in a generalized sense, we may call both Brutus and Cæsar subjects of this proposition: the fact that Brutus is grammatically subject and Cæsar object is logically irrelevant, since the same occurrence may be expressed in the words "Cæsar was killed by Brutus," where Cæsar is the grammatical subject. Thus in the simpler sort of proposition we shall have an attribute or relation holding of or between one, two or more "subjects" in the extended sense. (A relation may have more than two terms: *e.g.* "A gives B to C" is a relation of *three* terms.) Now it often happens that, on a closer scrutiny, the apparent subjects are found to be not really subjects, but to be capable of analysis; the only result of this, however, is that new subjects take their places. It also happens that the verb may grammatically be made subject: *e.g.* we may say, "Killing is a relation which holds between Brutus and Cæsar." But in such cases the grammar is misleading, and in a straightforward statement, following the rules that should guide philosophical grammar, Brutus and Cæsar will appear as the subjects and killing as the verb.

We are thus led to the conception of terms which, when they occur in propositions, can *only* occur as subjects, and never in any other way. This is part of the old scholastic definition of *substance*; but persistence through time, which belonged to that notion, forms no part of the notion with which we are concerned. We shall define "proper names" as those terms which can only occur as *subjects* in propositions (using "subject" in the extended sense just explained). We shall further define "individuals" or "particulars" as the objects that can be named by proper names. (It would be better to define them directly, rather than by means of the kind of symbols by which they are symbolized; but in order to do that we should have to plunge deeper into metaphysics than is desirable here.) It is, of course, possible that there is an endless regress: that whatever appears as a particular is really, on closer scrutiny, a class or some kind of complex. If this be the case, the axiom of infinity must of course be true. But if it be not the case, it must be theoretically possible for analysis to reach ultimate subjects, and it is these that give the meaning of "particulars" or "individuals." It is to the number of these that the axiom of infinity is assumed to apply. If it is true of them, it is true of classes of them, and classes of classes of them, and so on; similarly if it is false of them, it is false throughout this hierarchy. Hence it is natural to enunciate the axiom concerning them rather than concerning any other stage in the hierarchy. But whether the axiom is true or false, there seems no known method of discovering.

CHAPTER XIV

INCOMPATIBILITY AND THE THEORY OF DEDUCTION

We have now explored, somewhat hastily it is true, that part of the philosophy of mathematics which does not demand a critical examination of the idea of *class*. In the preceding chapter, however, we found ourselves confronted by problems which make such an examination imperative. Before we can undertake it, we must consider certain other parts of the philosophy of mathematics, which we have hitherto ignored. In a synthetic treatment, the parts which we shall now be concerned with come first: they are more fundamental than anything that we have discussed hitherto. Three topics will concern us before we reach the theory of classes, namely: (1) the theory of deduction, (2) propositional functions, (3) descriptions. Of these, the third is not logically presupposed in the theory of classes, but it is a simpler example of the *kind* of theory that is needed in dealing with classes. It is the first topic, the theory of deduction, that will concern us in the present chapter.

Mathematics is a deductive science: starting from certain premises, it arrives, by a strict process of deduction, at the various theorems which constitute it. It is true that, in the past, mathematical deductions were often greatly lacking in rigour; it is true also that perfect rigour is a scarcely attainable ideal. Nevertheless, in so far as rigour is lacking in a mathematical proof, the proof is defective; it is no defence to urge that common sense shows the result to be correct, for if we were to rely upon that, it would be better to dispense with argument altogether, rather than bring fallacy to the rescue of common sense. No appeal to common sense, or "intuition," or anything except strict deductive logic, ought to be needed in mathematics after the premises have been laid down.

Kant, having observed that the geometers of his day could not prove their theorems by unaided argument, but required an appeal to the figure, invented a theory of mathematical reasoning according to which the inference is never strictly logical, but always requires the support of what is called "intuition." The whole trend of modern mathematics, with its increased pursuit of rigour, has been against this Kantian theory. The things in the mathematics of Kant's day which cannot be *proved*, cannot be *known*—for example, the axiom of parallels. What can be known, in mathematics and by mathematical methods, is what can be deduced from pure logic. What else is to belong to human knowledge must be ascertained otherwise—empirically, through the senses or through experience in some form, but not *a priori*. The positive grounds for this thesis are to be found in *Principia Mathematica*, *passim*; a controversial defence of it is given in the *Principles of Mathematics*. We cannot here do more than refer the reader to those works, since the subject is too vast for hasty treatment. Meanwhile, we shall assume that all mathematics is deductive, and proceed to inquire as to what is involved in deduction.

In deduction, we have one or more propositions called *premises*, from which we infer a proposition called the *conclusion*. For our purposes, it will be convenient, when there are originally several premises, to amalgamate them into a single proposition, so as to be able to speak of *the* premise as well as of *the* conclusion. Thus we may regard deduction as a process by which we pass from knowledge of a certain proposition, the premise, to knowledge of a certain other proposition, the conclusion. But we shall not regard such a process as *logical* deduction unless it is *correct*, *i.e.* unless there is such a relation between premise and conclusion that we have a right to believe the conclusion if we know the premise to be true. It is this relation that is chiefly of interest in the logical theory of deduction.

In order to be able validly to infer the truth of a proposition, we must know that some other proposition is true, and that there is between the two a relation of the sort called "implication," *i.e.* that (as we say) the premise "implies" the conclusion. (We shall define this relation shortly.) Or we may know that a certain other proposition is false, and that there is a relation between the two of the sort called "disjunction," expressed by "p or q," [32] so that the knowledge that the one is false allows us to infer that the other is true. Again, what we wish to infer may be the *falsehood* of some proposition, not its truth. This may be inferred from the truth of another proposition, provided we know that the two are "incompatible," *i.e.* that if one is true, the other is false. It may also be inferred from the falsehood of another proposition, in just the same circumstances in which the truth of the other might have been inferred from the truth of the one; *i.e.* from the falsehood of *p* we may infer the falsehood of *q*, when *q* implies *p*. All these four are cases of inference. When our minds are fixed upon inference, it seems natural to take "implication" as the primitive fundamental relation, since this is the relation which must hold between *p* and *q* if we are to be able to infer the *truth* of *q* from the *truth* of *p*. But for technical reasons this is not the best primitive idea to choose. Before proceeding to primitive ideas and definitions, let us consider further the various functions of propositions suggested by the above-mentioned relations of propositions.

[32] We shall use the letters *p*, *q*, *r*, *s*, *t* to denote variable propositions.

The simplest of such functions is the negative, "not-*p*." This is that function of *p* which is true when *p* is false, and false when *p* is true. It is convenient to speak of the truth of a proposition, or its falsehood, as its "truth-value" [33]; *i.e. truth* is the "truth-value" of a true proposition, and *falsehood* of a false one. Thus not-*p* has the opposite truth-value to *p*.

[33] This term is due to Frege.

We may take next disjunction, "*p* or *q*." This is a function whose truth-value is truth when *p* is true and also when *q* is true, but is falsehood when both *p* and *q* are false.

Next we may take *conjunction*, "*p* and *q*." This has truth for its truth-value when *p* and *q* are both true; otherwise it has falsehood for its truth-value.

Take next *incompatibility*, *i.e.* "*p* and *q* are not both true." This is the negation of conjunction; it is also the disjunction of the negations of *p* and *q*, *i.e.* it is "not-*p* or not-*q*." Its truth-value is truth when *p* is false and likewise when *q* is false; its truth-value is falsehood when *p* and *q* are both true.

Last take *implication*, *i.e.* "*p* implies *q*," or "if *p*, then *q*." This is to be understood in the widest sense that will allow us to infer the truth of *q* if we know the truth of *p*. Thus we interpret it as meaning: "Unless *p* is false, *q* is true," or "either *p* is false or *q* is true." (The fact that "implies" is capable of other meanings does not concern us; this is the meaning which is convenient for us.) That is to say, "*p* implies *q*" is to mean "not-*p* or *q*": its truth-value is to be truth if *p* is false, likewise if *q* is true, and is to be falsehood if *p* is true and *q* is false.

We have thus five functions: negation, disjunction, conjunction, incompatibility, and implication. We might have added others, for example, joint falsehood, "not-*p* and not-*q*," but the above five will suffice. Negation differs from the other four in being a function of *one* proposition, whereas the others are functions of *two*. But all five agree in this, that their truth-value depends only upon that of the propositions which are their arguments. Given the truth or falsehood of *p*, or of *p* and *q* (as the case may be), we are given the truth or falsehood of the negation, disjunction, conjunction, incompatibility, or implication. A function of propositions which has this property is called a "truth-function."

The whole meaning of a truth-function is exhausted by the statement of the circumstances under which it is true or false. "Not-*p*," for example, is simply that function of *p* which is true when *p* is false, and false when *p* is true: there is no further meaning to be assigned to it. The same applies to "*p* or *q*" and the rest. It follows that two truth-functions which have the same truth-value for all values of the argument are indistinguishable. For example, "p and q" is the negation of "not-p or not-q" and *vice versa*; thus either of these may be *defined* as the negation of the other. There is no further meaning in a truth-function over and above the conditions under which it is true or false.

It is clear that the above five truth-functions are not all independent. We can define some of them in terms of others. There is no great difficulty in reducing the number to two; the two chosen in *Principia Mathematica* are negation and disjunction. Implication is then defined as "not-p or q"; incompatibility as "not-p or not-q"; conjunction as the negation of incompatibility. But it has been shown by Sheffer [34] that we can be content with *one* primitive idea for all five, and by Nicod [35] that this enables us to reduce the primitive propositions required in the theory of deduction to two non-formal principles and one formal one. For this purpose, we may take as our one indefinable either incompatibility or joint falsehood. We will choose the former.

[34] Trans. Am. Math. Soc., vol. xiv. pp. 481-488.
[35] Proc. Camb. Phil. Soc., vol. xix., *i.*, January 1917.

Our primitive idea, now, is a certain truth-function called "incompatibility," which we will denote by p/q. Negation can be at once defined as the incompatibility of a proposition with itself, *i.e.* "not-p" is defined as "p/p." Disjunction is the incompatibility of not-p and not-q, *i.e.* it is $(p/p)|(q/q)$. Implication is the incompatibility of p and not-q, *i.e.* $p|(q/q)$. Conjunction is the negation of incompatibility, *i.e.* it is $(p/q)|(p/q)$. Thus all our four other functions are defined in terms of incompatibility.

It is obvious that there is no limit to the manufacture of truth-functions, either by introducing more arguments or by repeating arguments. What we are concerned with is the connection of this subject with inference.

If we know that p is true and that p implies q, we can proceed to assert q. There is always unavoidably *something* psychological about inference: inference is a method by which we arrive at new knowledge, and what is not psychological about it is the relation which allows us to infer correctly; but the actual passage from the assertion of p to the assertion of q is a psychological process, and we must not seek to represent it in purely logical terms.

In mathematical practice, when we infer, we have always some expression containing variable propositions, say p and q, which is known, in virtue of its form, to be true for all values of p and q; we have also some other expression, part of the former, which is also known to be true for all values of p and q; and in virtue of the principles of inference, we are able to drop this part of our original expression, and assert what is left. This somewhat abstract account may be made clearer by a few examples.

Let us assume that we know the five formal principles of deduction enumerated in *Principia Mathematica*. (M. Nicod has reduced these to one, but as it is a complicated proposition, we will begin with the five.) These five propositions are as follows:—

(1) "p or p" implies p—*i.e.* if either p is true or p is true, then p is true.

(2) q implies "p or q"—*i.e.* the disjunction "p or q" is true when one of its alternatives is true.

(3) "p or q" implies "q or p." This would not be required if we had a theoretically more perfect notation, since in the conception of disjunction there is no order involved, so that "p or q" and "q or p" should be identical. But since our symbols, in any convenient form, inevitably introduce an order, we need suitable assumptions for showing that the order is irrelevant.

(4) If either p is true or "q or r" is true, then either q is true or "p or r" is true. (The twist in this proposition serves to increase its deductive power.)

(5) If q implies r, then "p or q" implies "p or r."

These are the *formal* principles of deduction employed in *Principia Mathematica.* A formal principle of deduction has a double use, and it is in order to make this clear that we have cited the above five propositions. It has a use as the premise of an inference, and a use as establishing the fact that the premise implies the conclusion. In the schema of an inference we have a proposition p, and a proposition "p implies q," from which we infer q. Now when we are concerned with the principles of deduction, our apparatus of primitive propositions has to yield both the p and the "p implies q" of our inferences. That is to say, our rules of deduction are to be used, not *only* as *rules*, which is their use for establishing "p implies q," but *also* as substantive premises, *i.e.* as the p of our schema. Suppose, for example, we wish to prove that if p implies q, then if q implies r it follows that p implies r. We have here a relation of three propositions which state implications. Put

p_1=p implies q, p_2=q implies r, and p_3=p implies r.

Then we have to prove that p1 implies that p2 implies p3. Now take the fifth of our above principles, substitute not-p for p, and remember that "not-p or q" is by definition the same as "p implies q." Thus our fifth principle yields:

> "If q implies r, then 'p implies q' implies 'p implies r,'" *i.e.* "p_2 implies that p_1 implies p_3." Call this proposition A.

But the fourth of our principles, when we substitute not-p, not-q, for p and q, and remember the definition of implication, becomes:

> "If p implies that q implies r, then q implies that p implies r."

Writing p_2 in place of p, p_1 in place of q, and p_3 in place of r, this becomes:

> "If p_2 implies that p_1 implies p_3, then p1 implies that p_2 implies p_3." Call this B.

Now we proved by means of our fifth principle that

> "p_2 implies that p_1 implies p_3," which was what we called A.

Thus we have here an instance of the schema of inference, since A represents the p of our scheme, and B represents the "p implies q." Hence we arrive at q, namely,

> "*p1* implies that p_2 implies p_3,"

which was the proposition to be proved. In this proof, the adaptation of our fifth principle, which yields A, occurs as a substantive premise; while the adaptation of our fourth principle, which yields B, is used to give the *form* of the inference. The formal and material employments of premises in the theory of deduction are closely intertwined, and it is not very important to keep them separated, provided we realize that they are in theory distinct.

The earliest method of arriving at new results from a premise is one which is illustrated in the above deduction, but which itself can hardly be called deduction. The primitive propositions, whatever they may be, are to be regarded as asserted for all possible values of the variable propositions p, q, r which occur in them. We may therefore substitute for (say) p any expression whose value is always a proposition, *e.g.* not-p, "s implies t," and so on. By means of such substitutions we really obtain sets of special cases of our original proposition, but from a practical point of view we obtain what are virtually new propositions. The legitimacy of substitutions of this kind has to be insured by means of a non-formal principle of inference. [36]

[36] No such principle is enunciated in *Principia Mathematica* or in M. Nicod'*s* article mentioned above. But this would seem to be an omission.

We may now state the one formal principle of inference to which M. Nicod has reduced the five given above. For this purpose we will first show how certain truth-functions can be defined in terms of incompatibility. We saw already that

$p|(q/q)$ means "p implies q."

We now observe that

$p|(q/r)$ means "p implies both q and r."

For this expression means "p is incompatible with the incompatibility of q and r," *i.e.* "p implies that q and r are not incompatible," *i.e.* "p implies that q and r are both true"—for, as we saw, the conjunction of q and r is the negation of their incompatibility.

Observe next that $t|(t/t)$ means "t implies itself." This is a particular case of $p|(q/q)$.

Let us write p for the negation of p; thus p/s will mean the negation of p/s, *i.e.* it will mean the conjunction of p and s. It follows that

$(s/q)|p/s$

expresses the incompatibility of s/q with the conjunction of p and s; in other words, it states that if p and s are both true, s/q is false, *i.e.* s and q are both true; in still simpler words, it states that p and s jointly imply s and q jointly.

Now, put $P=p|(q/r)$,
$\pi=t|(t/t)$
$Q=(s/q)|p/s$.

Then M. Nicod'*s* sole formal principle of deduction is

$P|\pi/Q$,

in other words, P implies both π and Q.

He employs in addition one non-formal principle belonging to the theory of types (which need not concern us), and one corresponding to the principle that, given p, and given that p implies q, we can assert q. This principle is:

"If $p|(r/q)$ is true, and p is true, then q is true." From this apparatus the whole theory of deduction follows, except in so far as we are concerned with deduction from or to the existence or the universal truth of "propositional functions," which we shall consider in the next chapter.

There is, if I am not mistaken, a certain confusion in the minds of some authors as to the relation, between propositions, in virtue of which an inference is valid. In order that it may be *valid* to infer q from p, it is only necessary that p should be true and that the proposition "not-p or q" should be true. Whenever this is the case, it is clear that q must be true. But inference will only in fact take place when the proposition "not-p or q" is *known* otherwise than through knowledge of not-p or knowledge of q. Whenever p is false, "not-p or q" is true, but is useless for inference, which requires that p should be true. Whenever q is already known to be true, "not-p or q" is of course also known to be true, but is again useless for inference, since q is already known, and therefore does not need to be inferred. In fact, inference only arises when "not-p or q" can be known without our knowing already which of the two alternatives it is that makes the disjunction true. Now, the circumstances under which this occurs are those in which certain relations of form exist between p and q. For example, we know that if r implies the negation of s, then s implies the negation of r. Between "r implies not-s" and "s implies not-r" there is a formal relation which enables us to *know* that the first implies the second, without having first to know that the first is false or to know that the second is true. It is under such circumstances that the relation of implication is practically useful for drawing inferences.

But this formal relation is only required in order that we may be able to *know* that either the premise is false or the conclusion is true. It is the truth of "not-p or q" that is required for the *validity* of the inference; what is required further is only required for the practical feasibility of the inference. Professor C. I. Lewis [37] has especially studied the narrower, formal relation which we may call "formal deducibility." He urges that the wider relation, that expressed by "not-p or q," should not be called "implication." That is, however, a matter of words. Provided our use of words is consistent, it matters little how we define them. The essential point of difference between the theory which I advocate and the theory advocated by Professor Lewis is this: He maintains that, when one proposition q is "formally deducible" from another p, the relation which we perceive between them is one which he calls "strict implication," which is not the relation expressed by "not-p or q" but a narrower relation, holding only when there are certain formal connections between p and q. I maintain that, whether or not there be such a relation as he speaks of, it is in any case one that mathematics does not need, and therefore one that, on general grounds of economy, ought not to be admitted into our apparatus of fundamental notions; that, whenever the relation of "formal deducibility" holds between two propositions, it is the case that we can see that either the first is false or the second true, and that nothing beyond this fact is necessary to be admitted into our premises; and that, finally, the reasons of detail which Professor Lewis adduces against the view which I advocate can all be met in detail, and depend for their plausibility upon a covert and unconscious assumption of the point of view which I reject. I conclude, therefore, that there is no need to admit as a fundamental notion any form of implication not expressible as a truth-function.

[37] See Mind, vol. xxi., 1912, pp. 522–531; and vol. xxiii., 1914, pp. 240-247.

CHAPTER XV

PROPOSITIONAL FUNCTIONS

When, in the preceding chapter, we were discussing propositions, we did not attempt to give a definition of the word "proposition." But although the word cannot be formally defined, it is necessary to say something as to its meaning, in order to avoid the very common confusion with "propositional functions," which are to be the topic of the present chapter.

We mean by a "proposition" primarily a form of words which expresses what is either true or false. I say "primarily," because I do not wish to exclude other than verbal symbols, or even mere thoughts if they have a symbolic character. But I think the word "proposition" should be limited to what may, in some sense, be called "symbols," and further to such symbols as give expression to truth and falsehood. Thus "two and two are four" and "two and two are five" will be propositions, and so will "Socrates is a man" and "Socrates is not a man." The statement: "Whatever numbers a and b may be, $(a+b)^2=a^2+2ab+b^2$" is a proposition; but the bare formula "$(a+b)2=a^2+2ab+b^2$" alone is not, since it asserts nothing definite unless we are further told, or led to suppose, that a and b are to have all possible values, or are to have such-and-such values. The former of these is tacitly assumed, as a rule, in the enunciation of mathematical formulæ, which thus become propositions; but if no such assumption were made, they would be "propositional functions." A "propositional function," in fact, is an expression containing one or more undetermined constituents, such that, when values are assigned to these constituents, the expression becomes a proposition. In other words, it is a function whose values are propositions. But this latter definition must be used with caution. A descriptive function, *e.g.* "the hardest proposition in A'*s* mathematical treatise," will not be a propositional function, although its values are propositions. But in such a case the propositions are only described: in a propositional function, the values must actually *enunciate* propositions.

Examples of propositional functions are easy to give: "x is human" is a propositional function; so long as x remains undetermined, it is neither true nor false, but when a value is assigned to x it becomes a true or false proposition. Any mathematical equation is a propositional function. So long as the variables have no definite value, the equation is merely an expression awaiting determination in order to become a true or false proposition. If it is an equation containing one variable, it becomes true when the variable is made equal to a root of the equation, otherwise it becomes false; but if it is an "identity" it will be true when the variable is any number. The equation to a curve in a plane or to a surface in space is a propositional function, true for values of the co-ordinates belonging to points on the curve or surface, false for other values. Expressions of traditional logic such as "all A is B" are propositional functions: A and B have to be determined as definite classes before such expressions become true or false.

The notion of "cases" or "instances" depends upon propositional functions. Consider, for example, the kind of process suggested by what is called "generalization," and let us take some very primitive example, say, "lightning is followed by thunder." We have a number of "instances" of this, *i.e.* a number of propositions such as: "this is a flash of lightning and is followed by thunder." What are these occurrences "instances" of? They are instances of the propositional function: "If x is a flash of lightning, x is followed by thunder." The process of generalization (with whose validity we are fortunately not concerned) consists in passing from a number of such instances to the *universal* truth of the propositional function: "If x is a flash of lightning, x is followed by thunder." It will be found that, in an analogous way, propositional functions are always involved whenever we talk of instances or cases or examples.

We do not need to ask, or attempt to answer, the question: "What is a propositional function?" A propositional function standing all alone may be taken to be a mere schema, a mere shell, an empty receptacle for meaning, not something already significant. We are concerned with propositional functions, broadly speaking, in two ways: first, as involved in the notions "true in all cases" and "true in some cases"; secondly, as involved in the theory of classes and relations. The second of these topics we will postpone to a later chapter; the first must occupy us now.

When we say that something is "always true" or "true in all cases," it is clear that the "something" involved cannot be a proposition. A proposition is just true or false, and there is an end of the matter. There are no instances or cases of "Socrates is a man" or "Napoleon died at St Helena." These are propositions, and it would be meaningless to speak of their being true "in all cases." This phrase is only applicable to propositional *functions*. Take, for example, the sort of thing that is often said when causation is being discussed. (We are not concerned with the truth or falsehood of what is said, but only with its logical analysis.) We are told that A is, in every instance, followed by B. Now if there are "instances" of A, A must be some general concept of which it is significant to say "x_1 is A," "x_2 is A," "x_3 is A," and so on, where x_1, x_2, x_3 are particulars which are not identical one with another. This applies, *e.g.*, to our previous case of lightning. We say that lightning (A) is followed by thunder (B). But the separate flashes are particulars, not identical, but sharing the common property of being lightning. The only way of expressing a common property generally is to say that a common property of a number of objects is a propositional function which becomes true when any one of these objects is taken as the value of the variable. In this case all the objects are "instances" of the truth of the propositional function—for a propositional function, though it cannot itself be true or false, is true in certain instances and false in certain others, unless it is "always true" or "always false." When, to return to our example, we say that A is in every instance followed by B, we mean that, whatever x may be, if x is an A, it is followed by a B; that is, we are asserting that a certain propositional function is "always true."

Sentences involving such words as "all," "every," "a," "the," "some" require propositional functions for their interpretation. The way in which propositional functions occur can be explained by means of two of the above words, namely, "all" and "some."

There are, in the last analysis, only two things that can be done with a propositional function: one is to assert that it is true in *all* cases, the other to assert that it is true in at least one case, or in *some* cases (as we shall say, assuming that there is to be no necessary implication of a plurality of cases). All the other uses of propositional functions can be reduced to these two. When we say that a propositional function is true "in all cases," or "always" (as we shall also say, without any temporal suggestion), we mean that all its values are true. If "φx" is the function, and a is the right sort of object to be an argument to "φx," then φa is to be true, however a may have been chosen. For example, "if a is human, a is mortal" is true whether a is human or not; in fact, every proposition of this form is true. Thus the propositional function "if x is human, x is mortal" is "always true," or "true in all cases." Or, again, the statement "there are no unicorns" is the same as the statement "the propositional function 'x is not a unicorn' is true in all cases." The assertions in the preceding chapter about propositions, *e.g.* "'*p* or *q*' implies '*q* or *p*,'" are really assertions that certain propositional functions are true in all cases. We do not assert the above principle, for example, as being true only of this or that particular *p* or *q*, but as being true of any *p* or *q* concerning which it can be made significantly. The condition that a function is to be *significant* for a given argument is the same as the condition that it shall have a value for that argument, either true or false. The study of the conditions of significance belongs to the doctrine of types, which we shall not pursue beyond the sketch given in the preceding chapter.

Not only the principles of deduction, but all the primitive propositions of logic, consist of assertions that certain propositional functions are always true. If this were not the case, they would have to mention particular things or concepts—Socrates, or redness, or east and west, or what not—and clearly it is not the province of logic to make assertions which are true concerning one such thing or concept but not concerning another. It is part of the definition of logic (but not the whole of its definition) that all its propositions are completely general, *i.e.* they all consist of the assertion that some propositional function containing no constant terms is always true. We shall return in our final chapter to the discussion of propositional functions containing no constant terms. For the present we will proceed to the other thing that is to be done with a propositional function, namely, the assertion that it is "sometimes true," *i.e.* true in at least one instance.

When we say "there are men," that means that the propositional function "*x* is a man" is sometimes true. When we say "some men are Greeks," that means that the propositional function "*x* is a man and a Greek" is sometimes true. When we say "cannibals still exist in Africa," that means that the propositional function "*x* is a cannibal now in Africa" is sometimes true, *i.e.* is true for some values of *x*. To say "there are at least *n* individuals in the world" is to say that the propositional function "α is a class of individuals and a member of the cardinal number *n*" is sometimes true, or, as we may say, is true for certain values of α. This form of expression is more convenient when it is necessary to indicate which is the variable constituent which we are taking as the argument to our propositional function. For example, the above propositional function, which we may shorten to "α is a class of *n* individuals," contains two variables, α and *n*. The axiom of infinity, in the language of propositional functions, is: "The propositional function 'if *n* is an inductive number, it is true for some values of α that α is a class of *n* individuals' is true for all possible values of *n*." Here there is a subordinate function, "α is a class of *n* individuals," which is said to be, in respect of α, *sometimes* true; and the assertion that this happens if *n* is an inductive number is said to be, in respect of *n*, *always* true.

The statement that a function φx is always true is the negation of the statement that not-φx is sometimes true, and the statement that φx is sometimes true is the negation of the statement that not-φx is always true. Thus the statement "all men are mortals" is the negation of the statement that the function "*x* is an immortal man" is sometimes true. And the statement "there are unicorns" is the negation of the statement that the function "*x* is not a unicorn" is always true. [38] We say that φx is "never true" or "always false" if not-φx is always true. We can, if we choose, take one of the pair "always," "sometimes" as a primitive idea, and define the other by means of the one and negation. Thus if we choose "sometimes" as our primitive idea, we can define: "'φx is always true' is to mean 'it is false that not-φx is sometimes true.'" But for reasons connected with the theory of types it seems more correct to take both "always" and "sometimes" as primitive ideas, and define by their means the negation of propositions in which they occur. That is to say, assuming that we have already defined (or adopted as a primitive idea) the negation of propositions of the type to which φx belongs, we define: "The negation of 'φx always' is 'not-φx sometimes'; and the negation of 'φx sometimes' is 'not-φx always.'" In like manner we can re-define disjunction and the other truth-functions, as applied to propositions containing apparent variables, in terms of the definitions and primitive ideas for propositions containing no apparent variables. Propositions containing no apparent variables are called "elementary propositions." From these we can mount up step by step, using such methods as have just been indicated, to the theory of truth-functions as applied to propositions containing one, two, three ... variables, or any number up to *n*, where *n* is any assigned finite number.[39]

[38] For linguistic reasons, to avoid suggesting either the plural or the singular, it is often convenient to say "φx is not always false" rather than "φx sometimes" or "φx is sometimes true."

[39] The method of deduction is given in *Principia Mathematica*, vol. *i.* *9.

The forms which are taken as simplest in traditional formal logic are really far from being so, and all involve the assertion of all values or some values of a compound propositional function. Take, to begin with, "all S is P." We will take it that S is defined by a propositional function φx, and P by a propositional function ψx. *E.g.*, if S is *men*, φx will be "*x* is human"; if P is *mortals*, ψx will be "there is a time at which *x* dies." Then "all S is P" means: "'φx implies ψx' is always true." It is to be observed that "all S is P" does not apply only to those terms that actually are S'*s*; it says something equally about terms which are not S'*s*. Suppose we come across an *x* of which we do not know whether it is an S or not; still, our statement "all S is P" tells us something about *x*, namely, that if *x* is an S, then *x* is a P. And this is every bit as true when *x* is not an S as when *x* is an S. If it were not equally true in both cases, the *reductio ad absurdum* would not be a valid method; for the essence of this method consists in using implications in cases where (as it afterwards turns out) the hypothesis is false. We may put the matter another way. In order to understand "all S is P," it is not necessary to be able to enumerate what terms are S'*s*; provided we know what is meant by being an S and what by being a P, we can understand completely what is actually affirmed by "all S is P," however little we may know of actual instances of either. This shows that it is not merely the actual terms that are S'*s* that are relevant in the statement "all S is P," but all the terms concerning which the supposition that they are S'*s* is significant, *i.e.* all the terms that are S'*s*, together with all the terms that are not S'*s*—*i.e.* the whole of the appropriate logical "type." What applies to statements about all applies also to statements about some. "There are men," *e.g.*, means that "*x* is human" is true for *some* values of *x*. Here *all* values of *x* (*i.e.* all values for which "*x* is human" is significant, whether true or false) are relevant, and not only those that in fact are human. (This becomes obvious if we consider how we could prove such a statement to be *false*.) Every assertion about "all" or "some" thus involves not only the arguments that make a certain function true, but all that make it significant, *i.e.* all for which it has a value at all, whether true or false.

We may now proceed with our interpretation of the traditional forms of the old-fashioned formal logic. We assume that S is those terms *x* for which φx is true, and P is those for which ψx is true. (As we shall see in a later chapter, all classes are derived in this way from propositional functions.) Then:

"All S is P" means "'φx implies ψx' is always true."
"Some S is P" means "'φx and ψx' is sometimes true."
"No S is P" means "'φx implies not-ψx' is always true."
"Some S is not P" means "'φx and not-ψx' is sometimes true."

It will be observed that the propositional functions which are here asserted for all or some values are not φx and ψx themselves, but truth-functions of φx and ψx for the same argument *x*. The easiest way to conceive of the sort of thing that is intended is to start not from φx and ψx in general, but from φa and ψa, where a is some constant. Suppose we are considering "all men are mortal": we will begin with

"If Socrates is human, Socrates is mortal,"

and then we will regard "Socrates" as replaced by a variable *x* wherever "Socrates" occurs. The object to be secured is that, although *x* remains a variable, without any definite value, yet it is to have the same value in "φx" as in "ψx" when we are asserting that "φx implies ψx" is always true. This requires that we shall start with a function whose values are such as "φa implies ψa," rather than with two separate functions φx and ψx; for if we start with two separate functions we can never secure that the *x*, while remaining undetermined, shall have the same value in both.

For brevity we say "φx always implies ψx" when we mean that "φx implies ψx" is always true. Propositions of the form "φx always implies ψx" are called "formal implications"; this name is given equally if there are several variables.

The above definitions show how far removed from the simplest forms are such propositions as "all S is P," with which traditional logic begins. It is typical of the lack of analysis involved that traditional logic treats "all S is P" as a proposition of the same form as "*x* is P"—*e.g.*, it treats "all men are mortal" as of the same form as "Socrates is mortal." As we have just seen, the first is of the form "φx always implies ψx," while the second is of the form "ψx." The emphatic separation of these two forms, which was effected by Peano and Frege, was a very vital advance in symbolic logic.

It will be seen that "all S is P" and "no S is P" do not really differ in form, except by the substitution of not-ψx for ψx, and that the same applies to "some S is P" and "some S is not P." It should also be observed that the traditional rules of conversion are faulty, if we adopt the view, which is the only technically tolerable one, that such propositions as "all S is P" do not involve the "existence" of S'*s*, *i.e.* do not require that there should be terms which are S'*s*. The above definitions lead to the result that, if φx is always false, *i.e.* if there are no S'*s*, then "all S is P" and "no S is P" will both be true, whatever P may be. For, according to the definition in the last chapter, "φx implies ψx" means "not-φx or ψx," which is always true if not-φx is always true. At the first moment, this result might lead the reader to desire different definitions, but a little practical experience soon shows that any different definitions would be inconvenient and would conceal the important ideas. The proposition "φx always implies ψx, and φx is sometimes true" is essentially composite, and it would be very awkward to give this as the definition of "all S is P," for then we should have no language left for "φx always implies ψx," which is needed a hundred times for once that the other is needed. But, with our definitions, "all S is P" does not imply "some S is P," since the first allows the non-existence of S and the second does not; thus conversion *per accidens* becomes invalid, and some moods of the syllogism are fallacious, *e.g.* Darapti: "All M is S, all M is P, therefore some S is P," which fails if there is no M.

The notion of "existence" has several forms, one of which will occupy us in the next chapter; but the fundamental form is that which is derived immediately from the notion of "sometimes true." We say that an argument a "satisfies" a function φx if φa is true; this is the same sense in which the roots of an equation are said to satisfy the equation. Now if φx is sometimes true, we may say there are *x's* for which it is true, or we may say "arguments satisfying φx *exist.*" This is the fundamental meaning of the word "existence." Other meanings are either derived from this, or embody mere confusion of thought. We may correctly say "men exist," meaning that "*x* is a man" is sometimes true. But if we make a pseudo-syllogism: "Men exist, Socrates is a man, therefore Socrates exists," we are talking nonsense, since "Socrates" is not, like "men," merely an undetermined argument to a given propositional function. The fallacy is closely analogous to that of the argument: "Men are numerous, Socrates is a man, therefore Socrates is numerous." In this case it is obvious that the conclusion is nonsensical, but in the case of existence it is not obvious, for reasons which will appear more fully in the next chapter. For the present let us merely note the fact that, though it is correct to say "men exist," it is incorrect, or rather meaningless, to ascribe existence to a given particular *x* who happens to be a man. Generally, "terms satisfying φx exist" means "φx is sometimes true"; but "a exists" (where a is a term satisfying φx) is a mere noise or shape, devoid of significance. It will be found that by bearing in mind this simple fallacy we can solve many ancient philosophical puzzles concerning the meaning of existence.

Another set of notions as to which philosophy has allowed itself to fall into hopeless confusions through not sufficiently separating propositions and propositional functions are the notions of "modality": *necessary, possible*, and *impossible*. (Sometimes *contingent* or *assertoric* is used instead of *possible*.) The traditional view was that, among true propositions, some were necessary, while others were merely contingent or assertoric; while among false propositions some were impossible, namely, those whose contradictories were necessary, while others merely happened not to be true. In fact, however, there was never any clear account of what was added to truth by the conception of necessity. In the case of propositional functions, the threefold division is obvious. If "φx" is an undetermined value of a certain propositional function, it will be *necessary* if the function is always true, possible if it is sometimes true, and *impossible* if it is never true. This sort of situation arises in regard to probability, for example. Suppose a ball *x* is drawn from a bag which contains a number of balls: if all the balls are white, "*x* is white" is necessary; if some are white, it is possible; if none, it is impossible. Here all that is *known* about *x* is that it satisfies a certain propositional function, namely, "*x* was a ball in the bag." This is a situation which is general in probability problems and not uncommon in practical life—*e.g.* when a person calls of whom we know nothing except that he brings a letter of introduction from our friend so-and-so. In all such cases, as in regard to modality in general, the propositional function is relevant. For clear thinking, in many very diverse directions, the habit of keeping propositional functions sharply separated from propositions is of the utmost importance, and the failure to do so in the past has been a disgrace to philosophy.

CHAPTER XVI

DESCRIPTIONS

We dealt in the preceding chapter with the words *all* and *some*; in this chapter we shall consider the word *the* in the singular, and in the next chapter we shall consider the word *the* in the plural. It may be thought excessive to devote two chapters to one word, but to the philosophical mathematician it is a word of very great importance: like Browning's Grammarian with the enclitic δε, I would give the doctrine of this word if I were "dead from the waist down" and not merely in a prison.

We have already had occasion to mention "descriptive functions," *i.e.* such expressions as "the father of x" or "the sine of x." These are to be defined by first defining "descriptions."

A "description" may be of two sorts, definite and indefinite (or ambiguous). An indefinite description is a phrase of the form "a so-and-so," and a definite description is a phrase of the form "the so-and-so" (in the singular). Let us begin with the former.

"Who did you meet?" "I met a man." "That is a very indefinite description." We are therefore not departing from usage in our terminology. Our question is: What do I really assert when I assert "I met a man"? Let us assume, for the moment, that my assertion is true, and that in fact I met Jones. It is clear that what I assert is *not* "I met Jones." I may say "I met a man, but it was not Jones"; in that case, though I lie, I do not contradict myself, as I should do if when I say I met a man I really mean that I met Jones. It is clear also that the person to whom I am speaking can understand what I say, even if he is a foreigner and has never heard of Jones.

But we may go further: not only Jones, but no actual man, enters into my statement. This becomes obvious when the statement is false, since then there is no more reason why Jones should be supposed to enter into the proposition than why anyone else should. Indeed the statement would remain significant, though it could not possibly be true, even if there were no man at all. "I met a unicorn" or "I met a sea-serpent" is a perfectly significant assertion, if we know what it would be to be a unicorn or a sea-serpent, *i.e.* what is the definition of these fabulous monsters. Thus it is only what we may call the *concept* that enters into the proposition. In the case of "unicorn," for example, there is only the concept: there is not also, somewhere among the shades, something unreal which may be called "a unicorn." Therefore, since it is significant (though false) to say "I met a unicorn," it is clear that this proposition, rightly analysed, does not contain a constituent "a unicorn," though it does contain the concept "unicorn."

The question of "unreality," which confronts us at this point, is a very important one. Misled by grammar, the great majority of those logicians who have dealt with this question have dealt with it on mistaken lines. They have regarded grammatical form as a surer guide in analysis than, in fact, it is. And they have not known what differences in grammatical form are important. "I met Jones" and "I met a man" would count traditionally as propositions of the same form, but in actual fact they are of quite different forms: the first names an actual person, Jones; while the second involves a propositional function, and becomes, when made explicit: "The function 'I met x and x is human' is sometimes true." (It will be remembered that we adopted the convention of using "sometimes" as not implying more than once.) This proposition is obviously not of the form "I met x," which accounts for the existence of the proposition "I met a unicorn" in spite of the fact that there is no such thing as "a unicorn."

For want of the apparatus of propositional functions, many logicians have been driven to the conclusion that there are unreal objects. It is argued, *e.g.* by Meinong, [40] that we can speak about "the golden mountain," "the round square," and so on; we can make true propositions of which these are the subjects; hence they must have some kind of logical being, since otherwise the propositions in which they occur would be meaningless. In such theories, it seems to me, there is a failure of that feeling for reality which ought to be preserved even in the most abstract studies. Logic, I should maintain, must no more admit a unicorn than zoology can; for logic is concerned with the real world just as truly as zoology, though with its more abstract and general features. To say that unicorns have an existence in heraldry, or in literature, or in imagination, is a most pitiful and paltry evasion. What exists in heraldry is not an animal, made of flesh and blood, moving and breathing of its own initiative. What exists is a picture, or a description in words. Similarly, to maintain that Hamlet, for example, exists in his own world, namely, in the world of Shakespeare's imagination, just as truly as (say) Napoleon existed in the ordinary world, is to say something deliberately confusing, or else confused to a degree which is scarcely credible. There is only one world, the "real" world: Shakespeare's imagination is part of it, and the thoughts that he had in writing Hamlet are real. So are the thoughts that we have in reading the play. But it is of the very essence of fiction that only the thoughts, feelings, etc., in Shakespeare and his readers are real, and that there is not, in addition to them, an objective Hamlet. When you have taken account of all the feelings roused by Napoleon in writers and readers of history, you have not touched the actual man; but in the case of Hamlet you have come to the end of him. If no one thought about Hamlet, there would be nothing left of him; if no one had thought about Napoleon, he would have soon seen to it that someone did. The sense of reality is vital in logic, and whoever juggles with it by pretending that Hamlet has another kind of reality is doing a disservice to thought. A robust sense of reality is very necessary in framing a correct analysis of propositions about unicorns, golden mountains, round squares, and other such pseudo-objects.

[40] *Untersuchungen zur Gegenstandstheorie und Psychologie*, 1904.

In obedience to the feeling of reality, we shall insist that, in the analysis of propositions, nothing "unreal" is to be admitted. But, after all, if there is nothing unreal, how, it may be asked, *could* we admit anything unreal? The reply is that, in dealing with propositions, we are dealing in the first instance with symbols, and if we attribute significance to groups of symbols which have no significance, we shall fall into the error of admitting unrealities, in the only sense in which this is possible, namely, as objects described. In the proposition "I met a unicorn," the whole four words together make a significant proposition, and the word "unicorn" by itself is significant, in just the same sense as the word "man." But the two words "a unicorn" do not form a subordinate group having a meaning of its own. Thus if we falsely attribute meaning to these two words, we find ourselves saddled with "a unicorn," and with the problem how there can be such a thing in a world where there are no unicorns. "A unicorn" is an indefinite description which describes nothing. It is not an indefinite description which describes something unreal. Such a proposition as "x is unreal" only has meaning when "x" is a description, definite or indefinite; in that case the proposition will be true if "x" is a description which describes nothing. But whether the description "x" describes something or describes nothing, it is in any case not a constituent of the proposition in which it occurs; like "a unicorn" just now, it is not a subordinate group having a meaning of its own. All this results from the fact that, when "*x*" is a description, "*x* is unreal" or "*x* does not exist" is not nonsense, but is always significant and sometimes true.

We may now proceed to define generally the meaning of propositions which contain ambiguous descriptions. Suppose we wish to make some statement about "a so-and-so," where "so-and-so'*s*" are those objects that have a certain property φ, *i.e.* those objects *x* for which the propositional function φx is true. (*E.g.* if we take "a man" as our instance of "a so-and-so," φx will be "*x* is human.") Let us now wish to assert the property ψ of "a so-and-so," *i.e.* we wish to assert that "a so-and-so" has that property which *x* has when ψx is true. (*E.g.* in the case of "I met a man," ψx will be "I met *x*.") Now the proposition that "a so-and-so" has the property ψ is not a proposition of the form "ψx." If it were, "a so-and-so" would have to be identical with *x* for a suitable *x*; and although (in a sense) this may be true in some cases, it is certainly not true in such a case as "a unicorn." It is just this fact, that the statement that a so-and-so has the property ψ is not of the form ψx, which makes it possible for "a so-and-so" to be, in a certain clearly definable sense, "unreal." The definition is as follows:—

The statement that "an object having the property φ has the property ψ"

means:

"The joint assertion of φx and ψx is not always false."

So far as logic goes, this is the same proposition as might be expressed by "some φ'*s* are ψ'*s*"; but rhetorically there is a difference, because in the one case there is a suggestion of singularity, and in the other case of plurality. This, however, is not the important point. The important point is that, when rightly analysed, propositions verbally about "a so-and-so" are found to contain no constituent represented by this phrase. And that is why such propositions can be significant even when there is no such thing as a so-and-so.

The definition of *existence*, as applied to ambiguous descriptions, results from what was said at the end of the preceding chapter. We say that "men exist" or "a man exists" if the propositional function "*x* is human" is sometimes true; and generally "a so-and-so" exists if "*x* is so-and-so" is sometimes true. We may put this in other language. The proposition "Socrates is a man" is no doubt *equivalent* to "Socrates is human," but it is not the very same proposition. The is of "Socrates is human" expresses the relation of subject and predicate; the is of "Socrates is a man" expresses identity. It is a disgrace to the human race that it has chosen to employ the same word "is" for these two entirely different ideas—a disgrace which a symbolic logical language of course remedies. The identity in "Socrates is a man" is identity between an object named (accepting "Socrates" as a name, subject to qualifications explained later) and an object ambiguously described. An object ambiguously described will "exist" when at least one such proposition is true, *i.e.* when there is at least one true proposition of the form "*x* is a so-and-so," where "*x*" is a name. It is characteristic of ambiguous (as opposed to definite) descriptions that there may be any number of true propositions of the above form—Socrates is a man, Plato is a man, etc. Thus "a man exists" follows from Socrates, or Plato, or anyone else. With definite descriptions, on the other hand, the corresponding form of proposition, namely,

"x is the so-and-so" (where "x" is a name), can only be true for one value of x at most. This brings us to the subject of definite descriptions, which are to be defined in a way analogous to that employed for ambiguous descriptions, but rather more complicated.

We come now to the main subject of the present chapter, namely, the definition of the word the (in the singular). One very important point about the definition of "a so-and-so" applies equally to "the so-and-so"; the definition to be sought is a definition of propositions in which this phrase occurs, not a definition of the phrase itself in isolation. In the case of "a so-and-so," this is fairly obvious: no one could suppose that "a man" was a definite object, which could be defined by itself. Socrates is a man, Plato is a man, Aristotle is a man, but we cannot infer that "a man" means the same as "Socrates" means and also the same as "Plato" means and also the same as "Aristotle" means, since these three names have different meanings. Nevertheless, when we have enumerated all the men in the world, there is nothing left of which we can say, "This is a man, and not only so, but it is the 'a man,' the quintessential entity that is just an indefinite man without being anybody in particular." It is of course quite clear that whatever there is in the world is definite: if it is a man it is one definite man and not any other. Thus there cannot be such an entity as "a man" to be found in the world, as opposed to specific men. And accordingly it is natural that we do not define "a man" itself, but only the propositions in which it occurs.

In the case of "the so-and-so" this is equally true, though at first sight less obvious. We may demonstrate that this must be the case, by a consideration of the difference between a name and a *definite description*. Take the proposition, "Scott is the author of *Waverley*." We have here a name, "Scott," and a description, "the author of *Waverley*," which are asserted to apply to the same person. The distinction between a name and all other symbols may be explained as follows:—

A name is a simple symbol whose meaning is something that can only occur as subject, *i.e.* something of the kind that, in Chapter XIII., we defined as an "individual" or a "particular." And a "simple" symbol is one which has no parts that are symbols. Thus "Scott" is a simple symbol, because, though it has parts (namely, separate letters), these parts are not symbols. On the other hand, "the author of *Waverley*" is not a simple symbol, because the separate words that compose the phrase are parts which are symbols. If, as may be the case, whatever *seems* to be an "individual" is really capable of further analysis, we shall have to content ourselves with what may be called "relative individuals," which will be terms that, throughout the context in question, are never analysed and never occur otherwise than as subjects. And in that case we shall have correspondingly to content ourselves with "relative names." From the standpoint of our present problem, namely, the definition of descriptions, this problem, whether these are absolute names or only relative names, may be ignored, since it concerns different stages in the hierarchy of "types," whereas we have to compare such couples as "Scott" and "the author of *Waverley*," which both apply to the same object, and do not raise the problem of types. We may, therefore, for the moment, treat names as capable of being absolute; nothing that we shall have to say will depend upon this assumption, but the wording may be a little shortened by it.

We have, then, two things to compare: (1) a name, which is a simple symbol, directly designating an individual which is its meaning, and having this meaning in its own right, independently of the meanings of all other words; (2) a description, which consists of several words, whose meanings are already fixed, and from which results whatever is to be taken as the "meaning" of the description.

A proposition containing a description is not identical with what that proposition becomes when a name is substituted, even if the name names the same object as the description describes. "Scott is the author of *Waverley*" is obviously a different proposition from "Scott is Scott": the first is a fact in literary history, the second a trivial truism. And if we put anyone other than Scott in place of "the author of *Waverley*," our proposition would become false, and would therefore certainly no longer be the same proposition. But, it may be said, our proposition is essentially of the same form as (say) "Scott is Sir Walter," in which two names are said to apply to the same person. The reply is that, if "Scott is Sir Walter" really means "the person named 'Scott' is the person named 'Sir Walter,'" then the names are being used as descriptions: *i.e.* the individual, instead of being named, is being described as the person having that name. This is a way in which names are frequently used in practice, and there will, as a rule, be nothing in the phraseology to show whether they are being used in this way or as names. When a name is used directly, merely to indicate what we are speaking about, it is no part of the *fact* asserted, or of the falsehood if our assertion happens to be false: it is merely part of the symbolism by which we express our thought. What we want to express is something which might (for example) be translated into a foreign language; it is something for which the actual words are a vehicle, but of which they are no part. On the other hand, when we make a proposition about "the person called 'Scott,'" the actual name "Scott" enters into what we are asserting, and not merely into the language used in making the assertion. Our proposition will now be a different one if we substitute "the person called 'Sir Walter.'" But so long as we are using names as names, whether we say "Scott" or whether we say "Sir Walter" is as irrelevant to what we are asserting as whether we speak English or French. Thus so long as names are used as names, "Scott is Sir Walter" is the same trivial proposition as "Scott is Scott." This completes the proof that "Scott is the author of *Waverley*" is not the same proposition as results from substituting a name for "the author of *Waverley*," no matter what name may be substituted.

When we use a variable, and speak of a propositional function, φx say, the process of applying general statements about φx to particular cases will consist in substituting a name for the letter "*x*," assuming that φ is a function which has individuals for its arguments. Suppose, for example, that φx is "always true"; let it be, say, the "law of identity," $x=x$. Then we may substitute for "*x*" any name we choose, and we shall obtain a true proposition. Assuming for the moment that "Socrates," "Plato," and "Aristotle" are names (a very rash assumption), we can infer from the law of identity that Socrates is Socrates, Plato is Plato, and Aristotle is Aristotle. But we shall commit a fallacy if we attempt to infer, without further premises, that the author of *Waverley* is the author of *Waverley*. This results from what we have just proved, that, if we substitute a name for "the author of *Waverley*" in a proposition, the proposition we obtain is a different one. That is to say, applying the result to our present case: If "*x*" is a name, "$x=x$" is not the same proposition as "the author of *Waverley* is the author of *Waverley*," no matter what name "*x*" may be. Thus from the fact that all propositions of the form "$x=x$" are true we cannot infer, without more ado, that the author of *Waverley* is the author of *Waverley*. In fact, propositions of the form "the so-and-so is the so-and-so" are not always true: it is necessary that the so-and-so should exist (a term which will be explained shortly). It is false that the present King of France is the present King of France, or that the round square is the round square. When we substitute a description for a name, propositional functions which are "always true" may become false, if the description describes nothing. There is no mystery in this as soon as we realize (what was proved in the preceding paragraph) that when we substitute a description the result is not a value of the propositional function in question.

We are now in a position to define propositions in which a definite description occurs. The only thing that distinguishes "*the* so-and-so" from "a so-and-so" is the implication of uniqueness. We cannot speak of "the inhabitant of London," because inhabiting London is an attribute which is not unique. We cannot speak about "the present King of France," because there is none; but we can speak about "the present King of England." Thus propositions about "the so-and-so" always imply the corresponding propositions about "a so-and-so," with the addendum that there is not more than one so-and-so. Such a proposition as "Scott is the author of *Waverley*" could not be true if *Waverley* had never been written, or if several people had written it; and no more could any other proposition resulting from a propositional function φx by the substitution of "the author of *Waverley*" for "*x*." We may say that "the author of *Waverley*" means "the value of *x* for which '*x* wrote Waverley' is true." Thus the proposition "the author of *Waverley* was Scotch," for example, involves:

(1) "*x* wrote *Waverley*" is not always false;
(2) "if *x* and *y* wrote *Waverley*, *x* and *y* are identical" is always true;
(3) "if *x* wrote *Waverley*, *x* was Scotch" is always true.

These three propositions, translated into ordinary language, state:

(1) at least one person wrote *Waverley*;
(2) at most one person wrote *Waverley*;
(3) whoever wrote *Waverley* was Scotch.

All these three are implied by "the author of *Waverley* was Scotch." Conversely, the three together (but no two of them) imply that the author of *Waverley* was Scotch. Hence the three together may be taken as defining what is meant by the proposition "the author of *Waverley* was Scotch."

We may somewhat simplify these three propositions. The first and second together are equivalent to: "There is a term *c* such that '*x* wrote *Waverley*' is true when *x* is *c* and is false when *x* is not *c*." In other words, "There is a term *c* such that '*x* wrote *Waverley*' is always equivalent to '*x* is *c*.'" (Two propositions are "equivalent" when both are true or both are false.) We have here, to begin with, two functions of *x*, "*x* wrote *Waverley*" and "*x* is *c*," and we form a function of *c* by considering the equivalence of these two functions of *x* for all values of *x*; we then proceed to assert that the resulting function of *c* is "sometimes true," *i.e.* that it is true for at least one value of *c*. (It obviously cannot be true for more than one value of *c*.) These two conditions together are defined as giving the meaning of "the author of *Waverley* exists."

We may now define "the term satisfying the function φx exists." This is the general form of which the above is a particular case. "The author of *Waverley*" is "the term satisfying the function '*x* wrote *Waverley*.'" And "the so-and-so" will always involve reference to some propositional function, namely, that which defines the property that makes a thing a so-and-so. Our definition is as follows:—

"The term satisfying the function φx exists" means:

"There is a term *c* such that φx is always equivalent to '*x* is *c*.'"

In order to define "the author of *Waverley* was Scotch," we have still to take account of the third of our three propositions, namely, "Whoever wrote *Waverley* was Scotch." This will be satisfied by merely adding that the c in question is to be Scotch. Thus "the author of *Waverley* was Scotch" is:

> "There is a term c such that (1) 'x wrote *Waverley*' is always equivalent to 'x is c,' (2) c is Scotch."

And generally: "the term satisfying φx satisfies ψx" is defined as meaning:

> "There is a term c such that (1) φx is always equivalent to 'x is c,' (2) ψc is true."

This is the definition of propositions in which descriptions occur.

It is possible to have much knowledge concerning a term described, *i.e.* to know many propositions concerning "the so-and-so," without actually knowing what the so-and-so is, *i.e.* without knowing any proposition of the form "x is the so-and-so," where "x" is a name. In a detective story propositions about "the man who did the deed" are accumulated, in the hope that ultimately they will suffice to demonstrate that it was A who did the deed. We may even go so far as to say that, in all such knowledge as can be expressed in words—with the exception of "this" and "that" and a few other words of which the meaning varies on different occasions—no names, in the strict sense, occur, but what seem like names are really descriptions. We may inquire significantly whether Homer existed, which we could not do if "Homer" were a name. The proposition "the so-and-so exists" is significant, whether true or false; but if a is the so-and-so (where "a" is a name), the words "a exists" are meaningless. It is only of descriptions —definite or indefinite—that existence can be significantly asserted; for, if "a" is a name, it *must* name something: what does not name anything is not a name, and therefore, if intended to be a name, is a symbol devoid of meaning, whereas a description, like "the present King of France," does not become incapable of occurring significantly merely on the ground that it describes nothing, the reason being that it is a *complex* symbol, of which the meaning is derived from that of its constituent symbols. And so, when we ask whether Homer existed, we are using the word "Homer" as an abbreviated description: we may replace it by (say) "the author of the *Iliad* and the *Odyssey*." The same considerations apply to almost all uses of what look like proper names.

When descriptions occur in propositions, it is necessary to distinguish what may be called "primary" and "secondary" occurrences. The abstract distinction is as follows. A description has a "primary" occurrence when the proposition in which it occurs results from substituting the description for "x" in some propositional function φx; a description has a "secondary" occurrence when the result of substituting the description for x in φx gives only *part* of the proposition concerned. An instance will make this clearer. Consider "the present King of France is bald." Here "the present King of France" has a primary occurrence, and the proposition is false. Every proposition in which a description which describes nothing has a primary occurrence is false. But now consider "the present King of France is not bald." This is ambiguous. If we are first to take "x is bald," then substitute "the present King of France" for "x," and then deny the result, the occurrence of "the present King of France" is secondary and our proposition is true; but if we are to take "x is not bald" and substitute "the present King of France" for "x," then "the present King of France" has a primary occurrence and the proposition is false. Confusion of primary and secondary occurrences is a ready source of fallacies where descriptions are concerned.

Descriptions occur in mathematics chiefly in the form of *descriptive functions*, *i.e.* "the term having the relation R to y," or "the R of y" as we may say, on the analogy of "the father of y" and similar phrases. To say "the father of y is rich," for example, is to say that the following propositional function of c: "c is rich, and 'x begat y' is always equivalent to 'x is c,'" is "sometimes true," *i.e.* is true for at least one value of c. It obviously cannot be true for more than one value.

The theory of descriptions, briefly outlined in the present chapter, is of the utmost importance both in logic and in theory of knowledge. But for purposes of mathematics, the more philosophical parts of the theory are not essential, and have therefore been omitted in the above account, which has confined itself to the barest mathematical requisites.

CHAPTER XVII

CLASSES

In the present chapter we shall be concerned with *the* in the plural: the inhabitants of London, the sons of rich men, and so on. In other words, we shall be concerned with *classes*. We saw in Chapter II. that a cardinal number is to be defined as a class of classes, and in Chapter III. that the number 1 is to be defined as the class of all unit classes, *i.e.* of all that have just one member, as we should say but for the vicious circle. Of course, when the number 1 is defined as the class of all unit classes, "unit classes" must be defined so as not to assume that we know what is meant by "one"; in fact, they are defined in a way closely analogous to that used for descriptions, namely: A class α is said to be a "unit" class if the propositional function "'x is an α' is always equivalent to 'x is c'" (regarded as a function of c) is not always false, *i.e.*, in more ordinary language, if there is a term c such that x will be a member of α when x is c but not otherwise. This gives us a definition of a unit class if we already know what a class is in general. Hitherto we have, in dealing with arithmetic, treated "class" as a primitive idea. But, for the reasons set forth in Chapter XIII., if for no others, we cannot accept "class" as a primitive idea. We must seek a definition on the same lines as the definition of descriptions, *i.e.* a definition which will assign a meaning to propositions in whose verbal or symbolic expression words or symbols apparently representing classes occur, but which will assign a meaning that altogether eliminates all mention of classes from a right analysis of such propositions. We shall then be able to say that the symbols for classes are mere conveniences, not representing objects called "classes," and that classes are in fact, like descriptions, logical fictions, or (as we say) "incomplete symbols."

The theory of classes is less complete than the theory of descriptions, and there are reasons (which we shall give in outline) for regarding the definition of classes that will be suggested as not finally satisfactory. Some further subtlety appears to be required; but the reasons for regarding the definition which will be offered as being approximately correct and on the right lines are overwhelming.

The first thing is to realize why classes cannot be regarded as part of the ultimate furniture of the world. It is difficult to explain precisely what one means by this statement, but one consequence which it implies may be used to elucidate its meaning. If we had a complete symbolic language, with a definition for everything definable, and an undefined symbol for everything indefinable, the undefined symbols in this language would represent symbolically what I mean by "the ultimate furniture of the world." I am maintaining that no symbols either for "class" in general or for particular classes would be included in this apparatus of undefined symbols. On the other hand, all the particular things there are in the world would have to have names which would be included among undefined symbols. We might try to avoid this conclusion by the use of descriptions. Take (say) "the last thing Cæsar saw before he died." This is a description of some particular; we might use it as (in one perfectly legitimate sense) a *definition* of that particular. But if "*a*" is a *name* for the same particular, a proposition in which "a" occurs is not (as we saw in the preceding chapter) identical with what this proposition becomes when for "*a*" we substitute "the last thing Cæsar saw before he died." If our language does not contain the name "*a*," or some other name for the same particular, we shall have no means of expressing the proposition which we expressed by means of "*a*" as opposed to the one that we expressed by means of the description. Thus descriptions would not enable a perfect language to dispense with names for all particulars. In this respect, we are maintaining, classes differ from particulars, and need not be represented by undefined symbols. Our first business is to give the reasons for this opinion.

We have already seen that classes cannot be regarded as a species of individuals, on account of the contradiction about classes which are not members of themselves (explained in Chapter XIII.), and because we can prove that the number of classes is greater than the number of individuals.

We cannot take classes in the *pure* extensional way as simply heaps or conglomerations. If we were to attempt to do that, we should find it impossible to understand how there can be such a class as the null-class, which has no members at all and cannot be regarded as a "heap"; we should also find it very hard to understand how it comes about that a class which has only one member is not identical with that one member. I do not mean to assert, or to deny, that there are such entities as "heaps." As a mathematical logician, I am not called upon to have an opinion on this point. All that I am maintaining is that, if there are such things as heaps, we cannot identify them with the classes composed of their constituents.

We shall come much nearer to a satisfactory theory if we try to identify classes with propositional functions. Every class, as we explained in Chapter II., is defined by some propositional function which is true of the members of the class and false of other things. But if a class can be defined by one propositional function, it can equally well be defined by any other which is true whenever the first is true and false whenever the first is false. For this reason the class cannot be identified with any one such propositional function rather than with any other—and given a propositional function, there are always many others which are true when it is true and false when it is false. We say that two propositional functions are "formally equivalent" when this happens. Two *propositions* are "equivalent" when both are true or both false; two propositional functions φx, ψx are "formally equivalent" when φx is always equivalent to ψx. It is the fact that there are other functions formally equivalent to a given function that makes it impossible to identify a class with a function; for we wish classes to be such that no two distinct classes have exactly the same members, and therefore two formally equivalent functions will have to determine the same class.

When we have decided that classes cannot be things of the same sort as their members, that they cannot be just heaps or aggregates, and also that they cannot be identified with propositional functions, it becomes very difficult to see what they can be, if they are to be more than symbolic fictions. And if we can find any way of dealing with them as symbolic fictions, we increase the logical security of our position, since we avoid the need of assuming that there are classes without being compelled to make the opposite assumption that there are no classes. We merely abstain from both assumptions. This is an example of Occam'*s* razor, namely, "entities are not to be multiplied without necessity." But when we refuse to assert that there are classes, we must not be supposed to be asserting dogmatically that there are none. We are merely agnostic as regards them: like Laplace, we can say, "*je n'ai pas besoin de cette hypothèse*."

Let us set forth the conditions that a symbol must fulfil if it is to serve as a class. I think the following conditions will be found necessary and sufficient:—

(1) Every propositional function must determine a class, consisting of those arguments for which the function is true. Given any proposition (true or false), say about Socrates, we can imagine Socrates replaced by Plato or Aristotle or a gorilla or the man in the moon or any other individual in the world. In general, some of these substitutions will give a true proposition and some a false one. The class determined will consist of all those substitutions that give a true one. Of course, we have still to decide what we mean by "all those which, etc." All that we are observing at present is that a class is rendered determinate by a propositional function, and that every propositional function determines an appropriate class.

(2) Two formally equivalent propositional functions must determine the same class, and two which are not formally equivalent must determine different classes. That is, a class is determined by its membership, and no two different classes can have the same membership. (If a class is determined by a function φx, we say that a is a "member" of the class if φa is true.)

(3) We must find some way of defining not only classes, but classes of classes. We saw in Chapter II. that cardinal numbers are to be defined as classes of classes. The ordinary phrase of elementary mathematics, "The combinations of *n* things *m* at a time" represents a class of classes, namely, the class of all classes of *m* terms that can be selected out of a given class of *n* terms. Without some symbolic method of dealing with classes of classes, mathematical logic would break down.

(4) It must under all circumstances be meaningless (not false) to suppose a class a member of itself or not a member of itself. This results from the contradiction which we discussed in Chapter XIII.

(5) Lastly—and this is the condition which is most difficult of fulfilment—it must be possible to make propositions about *all* the classes that are composed of individuals, or about *all* the classes that are composed of objects of any one logical "type." If this were not the case, many uses of classes would go astray—for example, mathematical induction. In defining the posterity of a given term, we need to be able to say that a member of the posterity belongs to *all* hereditary classes to which the given term belongs, and this requires the sort of totality that is in question. The reason there is a difficulty about this condition is that it can be proved to be impossible to speak of *all* the propositional functions that can have arguments of a given type.

We will, to begin with, ignore this last condition and the problems which it raises. The first two conditions may be taken together. They state that there is to be one class, no more and no less, for each group of formally equivalent propositional functions; *e.g.* the class of men is to be the same as that of featherless bipeds or rational animals or Yahoos or whatever other characteristic may be preferred for defining a human being. Now, when we say that two formally equivalent propositional functions may be not identical, although they define the same class, we may prove the truth of the assertion by pointing out that a statement may be true of the one function and false of the other; *e.g.* "I believe that all men are mortal" may be true, while "I believe that all rational animals are mortal" may be false, since I may believe falsely that the Phœnix is an immortal rational animal. Thus we are led to consider *statements about functions*, or (more correctly) *functions of functions*.

Some of the things that may be said about a function may be regarded as said about the class defined by the function, whereas others cannot. The statement "all men are mortal" involves the functions "x is human" and "x is mortal"; or, if we choose, we can say that it involves the classes *men* and *mortals*. We can interpret the statement in either way, because its truth-value is unchanged if we substitute for "x is human" or for "x is mortal" any formally equivalent function. But, as we have just seen, the statement "I believe that all men are mortal" cannot be regarded as being about the class determined by either function, because its truth-value may be changed by the substitution of a formally equivalent function (which leaves the class unchanged). We will call a statement involving a function φx an "extensional" function of the function φx, if it is like "all men are mortal," *i.e.* if its truth-value is unchanged by the substitution of any formally equivalent function; and when a function of a function is not extensional, we will call it "intensional," so that "I believe that all men are mortal" is an intensional function of "x is human" or "x is mortal." Thus *extensional* functions of a function φx may, for practical purposes, be regarded as functions of the class determined by φx, while *intensional* functions cannot be so regarded.

It is to be observed that all the *specific* functions of functions that we have occasion to introduce in mathematical logic are extensional. Thus, for example, the two fundamental functions of functions are: "φx is always true" and "φx is sometimes true." Each of these has its truth-value unchanged if any formally equivalent function is substituted for φx. In the language of classes, if α is the class determined by φx, "φx is always true" is equivalent to "everything is a member of α," and "φx is sometimes true" is equivalent to "α has members" or (better) "α has at least one member." Take, again, the condition, dealt with in the preceding chapter, for the existence of "the term satisfying φx." The condition is that there is a term c such that φx is always equivalent to "x is c." This is obviously extensional. It is equivalent to the assertion that the class defined by the function φx is a unit class, *i.e.* a class having one member; in other words, a class which is a member of 1.

Given a function of a function which may or may not be extensional, we can always derive from it a connected and certainly extensional function of the same function, by the following plan: Let our original function of a function be one which attributes to φx the property f; then consider the assertion "there is a function having the property f and formally equivalent to φx." This is an extensional function of φx; it is true when our original statement is true, and it is formally equivalent to the original function of φx if this original function is extensional; but when the original function is intensional, the new one is more often true than the old one. For example, consider again "I believe that all men are mortal," regarded as a function of "x is human." The derived extensional function is: "There is a function formally equivalent to 'x is human' and such that I believe that whatever satisfies it is mortal." This remains true when we substitute "x is a rational animal" for "x is human," even if I believe falsely that the Phœnix is rational and immortal.

We give the name of "derived extensional function" to the function constructed as above, namely, to the function: "There is a function having the property f and formally equivalent to φx," where the original function was "the function φx has the property *f*."

We may regard the derived extensional function as having for its argument the class determined by the function φx, and as asserting *f* of this class. This may be taken as the definition of a proposition about a class. *I.e.* we may define:

To assert that "the class determined by the function φx has the property *f*" is to assert that φx satisfies the extensional function derived from *f*.

This gives a meaning to any statement about a class which can be made significantly about a function; and it will be found that technically it yields the results which are required in order to make a theory symbolically satisfactory. [41]

[41] See *Principia Mathematica*, vol. *i*. pp. 75–84 and *20.

What we have said just now as regards the definition of classes is sufficient to satisfy our first four conditions. The way in which it secures the third and fourth, namely, the possibility of classes of classes, and the impossibility of a class being or not being a member of itself, is somewhat technical; it is explained in *Principia Mathematica*, but may be taken for granted here. It results that, but for our fifth condition, we might regard our task as completed. But this condition—at once the most important and the most difficult—is not fulfilled in virtue of anything we have said as yet. The difficulty is connected with the theory of types, and must be briefly discussed. [42]

[42] The reader who desires a fuller discussion should consult *Principia Mathematica*, Introduction, chap. ii.; also *12.

We saw in Chapter XIII. that there is a hierarchy of logical types, and that it is a fallacy to allow an object belonging to one of these to be substituted for an object belonging to another. Now it is not difficult to show that the various functions which can take a given object a as argument are not all of one type. Let us call them all a-functions. We may take first those among them which do not involve reference to any collection of functions; these we will call "predicative a-functions." If we now proceed to functions involving reference to the totality of predicative a-functions, we shall incur a fallacy if we regard these as of the same type as the predicative *a*-functions. Take such an every-day statement as "*a* is a typical Frenchman." How shall we define a "typical Frenchman"? We may define him as one "possessing all qualities that are possessed by most Frenchmen." But unless we confine "all qualities" to such as do not involve a reference to any totality of qualities, we shall have to observe that most Frenchmen are *not* typical in the above sense, and therefore the definition shows that to be not typical is essential to a typical Frenchman. This is not a logical contradiction, since there is no reason why there should be any typical Frenchmen; but it illustrates the need for separating off qualities that involve reference to a totality of qualities from those that do not.

Whenever, by statements about "all" or "some" of the values that a variable can significantly take, we generate a new object, this new object must not be among the values which our previous variable could take, since, if it were, the totality of values over which the variable could range would only be definable in terms of itself, and we should be involved in a vicious circle. For example, if I say "Napoleon had all the qualities that make a great general," I must define "qualities" in such a way that it will not include what I am now saying, *i.e.* "having all the qualities that make a great general" must not be itself a quality in the sense supposed. This is fairly obvious, and is the principle which leads to the theory of types by which vicious-circle paradoxes are avoided. As applied to *a*-functions, we may suppose that "qualities" is to mean "predicative functions." Then when I say "Napoleon had all the qualities, etc.," I mean "Napoleon satisfied all the predicative functions, etc." This statement attributes a property to Napoleon, but not a predicative property; thus we escape the vicious circle. But wherever "all functions which" occurs, the functions in question must be limited to one type if a vicious circle is to be avoided; and, as Napoleon and the typical Frenchman have shown, the type is not rendered determinate by that of the argument. It would require a much fuller discussion to set forth this point fully, but what has been said may suffice to make it clear that the functions which can take a given argument are of an infinite series of types. We could, by various technical devices, construct a variable which would run through the first n of these types, where n is finite, but we cannot construct a variable which will run through them all, and, if we could, that mere fact would at once generate a new type of function with the same arguments, and would set the whole process going again.

We call predicative a-functions the *first* type of *a*-functions; a-functions involving reference to the totality of the first type we call the *second* type; and so on. No variable *a*-function can run through all these different types: it must stop short at some definite one.

These considerations are relevant to our definition of the derived extensional function. We there spoke of "a function formally equivalent to φx." It is necessary to decide upon the type of our function. Any decision will do, but some decision is unavoidable. Let us call the supposed formally equivalent function ψ. Then ψ appears as a variable, and must be of some determinate type. All that we know necessarily about the type of φ is that it takes arguments of a given type—that it is (say) an *a*-function. But this, as we have just seen, does not determine its type. If we are to be able (as our fifth requisite demands) to deal with all classes whose members are of the same type as a, we must be able to define all such classes by means of functions of some one type; that is to say, there must be some type of a-function, say the nth, such that any a-function is formally equivalent to some a-function of the nth type. If this is the case, then any extensional function which holds of all a-functions of the nth type will hold of any *a*-function whatever. It is chiefly as a technical means of embodying an assumption leading to this result that classes are useful. The assumption is called the "axiom of reducibility," and may be stated as follows:—

"There is a type (τ say) of *a*-functions such that, given any a-function, it is formally equivalent to some function of the type in question."

If this axiom is assumed, we use functions of this type in defining our associated extensional function. Statements about all a-classes (*i.e.* all classes defined by a-functions) can be reduced to statements about all *a*-functions of the type τ. So long as only extensional functions of functions are involved, this gives us in practice results which would otherwise have required the impossible notion of "all *a*-functions." One particular region where this is vital is mathematical induction.

The axiom of reducibility involves all that is really essential in the theory of classes. It is therefore worthwhile to ask whether there is any reason to suppose it true.

This axiom, like the multiplicative axiom and the axiom of infinity, is necessary for certain results, but not for the bare existence of deductive reasoning. The theory of deduction, as explained in Chapter XIV., and the laws for propositions involving "all" and "some," are of the very texture of mathematical reasoning: without them, or something like them, we should not merely not obtain the same results, but we should not obtain any results at all. We cannot use them as hypotheses, and deduce hypothetical consequences, for they are rules of deduction as well as premises. They must be absolutely true, or else what we deduce according to them does not even follow from the premises. On the other hand, the axiom of reducibility, like our two previous mathematical axioms, could perfectly well be stated as an hypothesis whenever it is used, instead of being assumed to be actually true. We can deduce its consequences hypothetically; we can also deduce the consequences of supposing it false. It is therefore only convenient, not necessary. And in view of the complication of the theory of types, and of the uncertainty of all except its most general principles, it is impossible as yet to say whether there may not be some way of dispensing with the axiom of reducibility altogether. However, assuming the correctness of the theory outlined above, what can we say as to the truth or falsehood of the axiom?

The axiom, we may observe, is a generalized form of Leibniz'*s* identity of indiscernibles. Leibniz assumed, as a logical principle, that two different subjects must differ as to predicates. Now predicates are only some among what we called "predicative functions," which will include also relations to given terms, and various properties not to be reckoned as predicates. Thus Leibniz'*s* assumption is a much stricter and narrower one than ours. (Not, of course, according to *his* logic, which regarded *all* propositions as reducible to the subject-predicate form.) But there is no good reason for believing his form, so far as I can see. There might quite well, as a matter of abstract logical possibility, be two things which had exactly the same predicates, in the narrow sense in which we have been using the word "predicate." How does our axiom look when we pass beyond predicates in this narrow sense? In the actual world there seems no way of doubting its empirical truth as regards particulars, owing to spatio-temporal differentiation: no two particulars have exactly the same spatial and temporal relations to all other particulars. But this is, as it were, an accident, a fact about the world in which we happen to find ourselves. Pure logic, and pure mathematics (which is the same thing), aims at being true, in Leibnizian phraseology, in all possible worlds, not only in this higgledy-piggledy job-lot of a world in which chance has imprisoned us. There is a certain lordliness which the logician should preserve: he must not condescend to derive arguments from the things he sees about him.

Viewed from this strictly logical point of view, I do not see any reason to believe that the axiom of reducibility is logically necessary, which is what would be meant by saying that it is true in all possible worlds. The admission of this axiom into a system of logic is therefore a defect, even if the axiom is empirically true. It is for this reason that the theory of classes cannot be regarded as being as complete as the theory of descriptions. There is need of further work on the theory of types, in the hope of arriving at a doctrine of classes which does not require such a dubious assumption. But it is reasonable to regard the theory outlined in the present chapter as right in its main lines, *i.e.* in its reduction of propositions nominally about classes to propositions about their defining functions. The avoidance of classes as entities by this method must, it would seem, be sound in principle, however the detail may still require adjustment. It is because this seems indubitable that we have included the theory of classes, in spite of our desire to exclude, as far as possible, whatever seemed open to serious doubt.

The theory of classes, as above outlined, reduces itself to one axiom and one definition. For the sake of definiteness, we will here repeat them. The axiom is:

There is a type τ such that if φ is a function which can take a given object a as argument, then there is a function ψ of the type τ which is formally equivalent to φ.

The definition is:

If φ is a function which can take a given object a as argument, and τ the type mentioned in the above axiom, then to say that the class determined by φ has the property f is to say that there is a function of type τ, formally equivalent to φ, and having the property f.

CHAPTER XVIII

MATHEMATICS AND LOGIC

Mathematics and logic, historically speaking, have been entirely distinct studies. Mathematics has been connected with science, logic with Greek. But both have developed in modern times: logic has become more mathematical and mathematics has become more logical. The consequence is that it has now become wholly impossible to draw a line between the two; in fact, the two are one. They differ as boy and man: logic is the youth of mathematics and mathematics is the manhood of logic. This view is resented by logicians who, having spent their time in the study of classical texts, are incapable of following a piece of symbolic reasoning, and by mathematicians who have learnt a technique without troubling to inquire into its meaning or justification. Both types are now fortunately growing rarer. So much of modern mathematical work is obviously on the border-line of logic, so much of modern logic is symbolic and formal, that the very close relationship of logic and mathematics has become obvious to every instructed student. The proof of their identity is, of course, a matter of detail: starting with premises which would be universally admitted to belong to logic, and arriving by deduction at results which as obviously belong to mathematics, we find that there is no point at which a sharp line can be drawn, with logic to the left and mathematics to the right. If there are still those who do not admit the identity of logic and mathematics, we may challenge them to indicate at what point, in the successive definitions and deductions of *Principia Mathematica*, they consider that logic ends and mathematics begins. It will then be obvious that any answer must be quite arbitrary.

In the earlier chapters of this book, starting from the natural numbers, we have first defined "cardinal number" and shown how to generalize the conception of number, and have then analysed the conceptions involved in the definition, until we found ourselves dealing with the fundamentals of logic. In a synthetic, deductive treatment these fundamentals come first, and the natural numbers are only reached after a long journey. Such treatment, though formally more correct than that which we have adopted, is more difficult for the reader, because the ultimate logical concepts and propositions with which it starts are remote and unfamiliar as compared with the natural numbers. Also they represent the present frontier of knowledge, beyond which is the still unknown; and the dominion of knowledge over them is not as yet very secure.

It used to be said that mathematics is the science of "quantity." "Quantity" is a vague word, but for the sake of argument we may replace it by the word "number." The statement that mathematics is the science of number would be untrue in two different ways. On the one hand, there are recognised branches of mathematics which have nothing to do with number—all geometry that does not use co-ordinates or measurement, for example: projective and descriptive geometry, down to the point at which co-ordinates are introduced, does not have to do with number, or even with quantity in the sense of *greater* and *less*. On the other hand, through the definition of cardinals, through the theory of induction and ancestral relations, through the general theory of series, and through the definitions of the arithmetical operations, it has become possible to generalize much that used to be proved only in connection with numbers. The result is that what was formerly the single study of Arithmetic has now become divided into a number of separate studies, no one of which is specially concerned with numbers. The most elementary properties of numbers are concerned with one-one relations, and similarity between classes. Addition is concerned with the construction of mutually exclusive classes respectively similar to a set of classes which are not known to be mutually exclusive. Multiplication is merged in the theory of "selections," *i.e.* of a certain kind of one-many relations. Finitude is merged in the general study of ancestral relations, which yields the whole theory of mathematical induction. The ordinal properties of the various kinds of number-series, and the elements of the theory of continuity of functions and the limits of functions, can be generalized so as no longer to involve any essential reference to numbers. It is a principle, in all formal reasoning, to generalize to the utmost, since we thereby secure that a given process of deduction shall have more widely applicable results; we are, therefore, in thus generalizing the reasoning of arithmetic, merely following a precept which is universally admitted in mathematics. And in thus generalizing we have, in effect, created a set of new deductive systems, in which traditional arithmetic is at once dissolved and enlarged; but whether any one of these new deductive systems—for example, the theory of selections—is to be said to belong to logic or to arithmetic is entirely arbitrary, and incapable of being decided rationally.

We are thus brought face to face with the question: What is this subject, which may be called indifferently either mathematics or logic? Is there any way in which we can define it?

Certain characteristics of the subject are clear. To begin with, we do not, in this subject, deal with particular things or particular properties: we deal formally with what can be said about *any* thing or *any* property. We are prepared to say that one and one are two, but not that Socrates and Plato are two, because, in our capacity of logicians or pure mathematicians, we have never heard of Socrates and Plato. A world in which there were no such individuals would still be a world in which one and one are two. It is not open to us, as pure mathematicians or logicians, to mention anything at all, because, if we do so, we introduce something irrelevant and not formal. We may make this clear by applying it to the case of the syllogism. Traditional logic says: "All men are mortal, Socrates is a man, therefore Socrates is mortal." Now it is clear that what we *mean* to assert, to begin with, is only that the premises imply the conclusion, not that premises and conclusion are actually true; even the most traditional logic points out that the actual truth of the premises is irrelevant to logic. Thus the first change to be made in the above traditional syllogism is to state it in the form: "If all men are mortal and Socrates is a man, then Socrates is mortal." We may now observe that it is intended to convey that this argument is valid in virtue of its *form*, not in virtue of the particular terms occurring in it. If we had omitted "Socrates is a man" from our premises, we should have had a non-formal argument, only admissible because Socrates is in fact a man; in that case we could not have generalized the argument. But when, as above, the argument is *formal*, nothing depends upon the terms that occur in it. Thus we may substitute α for *men*, β for *mortals*, and x for Socrates, where α and β are any classes whatever, and x is any individual. We then arrive at the statement: "No matter what possible values x and α and β may have, if all α's are β's and x is an α, then x is a β"; in other words, "the propositional function 'if all α's are β's and x is an α, then x is a β' is always true." Here at last we have a proposition of logic—the one which is only *suggested* by the traditional statement about Socrates and men and mortals.

It is clear that, if *formal* reasoning is what we are aiming at, we shall always arrive ultimately at statements like the above, in which no actual things or properties are mentioned; this will happen through the mere desire not to waste our time proving in a particular case what can be proved generally. It would be ridiculous to go through a long argument about Socrates, and then go through precisely the same argument again about Plato. If our argument is one (say) which holds of all men, we shall prove it concerning "x," with the hypothesis "if x is a man." With this hypothesis, the argument will retain its hypothetical validity even when x is not a man. But now we shall find that our argument would still be valid if, instead of supposing x to be a man, we were to suppose him to be a monkey or a goose or a Prime Minister. We shall therefore not waste our time taking as our premise "x is a man" but shall take "x is an α," where α is any class of individuals, or "φx" where φ is any propositional function of some assigned type. Thus the absence of all mention of particular things or properties in logic or pure mathematics is a necessary result of the fact that this study is, as we say, "purely formal."

At this point we find ourselves faced with a problem which is easier to state than to solve. The problem is: "What are the constituents of a logical proposition?" I do not know the answer, but I propose to explain how the problem arises.

Take (say) the proposition "Socrates was before Aristotle." Here it seems obvious that we have a relation between two terms, and that the constituents of the proposition (as well as of the corresponding fact) are simply the two terms and the relation, *i.e.* Socrates, Aristotle, and *before*. (I ignore the fact that Socrates and Aristotle are not simple; also the fact that what appear to be their names are really truncated descriptions. Neither of these facts is relevant to the present issue.) We may represent the general form of such propositions by "x R y," which may be read "x has the relation R to y." This general form may occur in logical propositions, but no particular instance of it can occur. Are we to infer that the general form itself is a constituent of such logical propositions?

Given a proposition, such as "Socrates is before Aristotle," we have certain constituents and also a certain form. But the form is not itself a new constituent; if it were, we should need a new form to embrace both it and the other constituents. We can, in fact, turn *all* the constituents of a proposition into variables, while keeping the form unchanged. This is what we do when we use such a schema as "x R y," which stands for any one of a certain class of propositions, namely, those asserting relations between two terms. We can proceed to general assertions, such as "x R y is sometimes true"—*i.e.* there are cases where dual relations hold. This assertion will belong to logic (or mathematics) in the sense in which we are using the word. But in this assertion we do not mention any particular things or particular relations; no particular things or relations can ever enter into a proposition of pure logic. We are left with pure *forms* as the only possible constituents of logical propositions.

I do not wish to assert positively that pure forms—*e.g.* the form "x R y"—do actually enter into propositions of the kind we are considering. The question of the analysis of such propositions is a difficult one, with conflicting considerations on the one side and on the other. We cannot embark upon this question now, but we may accept, as a first approximation, the view that *forms* are what enter into logical propositions as their constituents. And we may explain (though not formally define) what we mean by the "form" of a proposition as follows:—

The "form" of a proposition is that, in it, that remains unchanged when every constituent of the proposition is replaced by another.

Thus "Socrates is earlier than Aristotle" has the same form as "Napoleon is greater than Wellington," though every constituent of the two propositions is different.

We may thus lay down, as a necessary (though not sufficient) characteristic of logical or mathematical propositions, that they are to be such as can be obtained from a proposition containing no variables (*i.e.* no such words as *all, some, a, the*, etc.) by turning every constituent into a variable and asserting that the result is always true or sometimes true, or that it is always true in respect of some of the variables that the result is sometimes true in respect of the others, or any variant of these forms. And another way of stating the same thing is to say that logic (or mathematics) is concerned only with *forms*, and is concerned with them only in the way of stating that they are always or sometimes true—with all the permutations of "always" and "sometimes" that may occur.

There are in every language some words whose sole function is to indicate form. These words, broadly speaking, are commonest in languages having fewest inflections. Take "Socrates is human." Here "is" is not a constituent of the proposition, but merely indicates the subject-predicate form. Similarly in "Socrates is earlier than Aristotle," "is" and "than" merely indicate form; the proposition is the same as "Socrates precedes Aristotle," in which these words have disappeared and the form is otherwise indicated. Form, as a rule, can be indicated otherwise than by specific words: the order of the words can do most of what is wanted. But this principle must not be pressed. For example, it is difficult to see how we could conveniently express molecular forms of propositions (*i.e.* what we call "truth-functions") without any word at all. We saw in Chapter XIV. that one word or symbol is enough for this purpose, namely, a word or symbol expressing *incompatibility*. But without even one we should find ourselves in difficulties. This, however, is not the point that is important for our present purpose. What is important for us is to observe that form may be the one concern of a general proposition, even when no word or symbol in that proposition designates the form. If we wish to speak about the form itself, we must have a word for it; but if, as in mathematics, we wish to speak about all propositions that have the form, a word for the form will usually be found not indispensable; probably in theory it is *never* indispensable.

Assuming—as I think we may—that the forms of propositions *can* be represented by the forms of the propositions in which they are expressed without any special words for forms, we should arrive at a language in which everything formal belonged to syntax and not to vocabulary. In such a language we could express *all* the propositions of mathematics even if we did not know one single word of the language. The language of mathematical logic, if it were perfected, would be such a language. We should have symbols for variables, such as "x" and "R" and "y," arranged in various ways; and the way of arrangement would indicate that something was being said to be true of all values or some values of the variables. We should not need to know any words, because they would only be needed for giving values to the variables, which is the business of the applied mathematician, not of the pure mathematician or logician. It is one of the marks of a proposition of logic that, given a suitable language, such a proposition can be asserted in such a language by a person who knows the syntax without knowing a single word of the vocabulary.

But, after all, there are words that express form, such as "is" and "than." And in every symbolism hitherto invented for mathematical logic there are symbols having constant formal meanings. We may take as an example the symbol for incompatibility which is employed in building up truth-functions. Such words or symbols may occur in logic. The question is: How are we to define them?

Such words or symbols express what are called "logical constants." Logical constants may be defined exactly as we defined forms; in fact, they are in essence the same thing. A fundamental logical constant will be that which is in common among a number of propositions, any one of which can result from any other by substitution of terms one for another. For example, "Napoleon is greater than Wellington" results from "Socrates is earlier than Aristotle" by the substitution of "Napoleon" for "Socrates," "Wellington" for "Aristotle," and "greater" for "earlier." Some propositions can be obtained in this way from the prototype "Socrates is earlier than Aristotle" and some cannot; those that can are those that are of the form "x R y," *i.e.* express dual relations. We cannot obtain from the above prototype by term-for-term substitution such propositions as "Socrates is human" or "the Athenians gave the hemlock to Socrates," because the first is of the subject-predicate form and the second expresses a three-term relation. If we are to have any words in our pure logical language, they must be such as express "logical constants," and "logical constants" will always either be, or be derived from, what is in common among a group of propositions derivable from each other, in the above manner, by term-for-term substitution. And this which is in common is what we call "form."

In this sense all the "constants" that occur in pure mathematics are logical constants. The number 1, for example, is derivative from propositions of the form: "There is a term c such that φx is true when, and only when, x is c." This is a function of φ, and various different propositions result from giving different values to φ. We may (with a little omission of intermediate steps not relevant to our present purpose) take the above function of φ as what is meant by "the class determined by φ is a unit class" or "the class determined by φ is a member of 1" (1 being a class of classes). In this way, propositions in which 1 occurs acquire a meaning which is derived from a certain constant logical form. And the same will be found to be the case with all mathematical constants: all are logical constants, or symbolic abbreviations whose full use in a proper context is defined by means of logical constants.

But although all logical (or mathematical) propositions can be expressed wholly in terms of logical constants together with variables, it is not the case that, conversely, all propositions that can be expressed in this way are logical. We have found so far a necessary but not a sufficient criterion of mathematical propositions. We have sufficiently defined the character of the primitive *ideas* in terms of which all the ideas of mathematics can be *defined*, but not of the primitive propositions from which all the *propositions* of mathematics can be *deduced*. This is a more difficult matter, as to which it is not yet known what the full answer is.

We may take the axiom of infinity as an example of a proposition which, though it can be enunciated in logical terms, cannot be asserted by logic to be true. All the propositions of logic have a characteristic which used to be expressed by saying that they were analytic, or that their contradictories were self-contradictory. This mode of statement, however, is not satisfactory. The law of contradiction is merely one among logical propositions; it has no special pre-eminence; and the proof that the contradictory of some proposition is self-contradictory is likely to require other principles of deduction besides the law of contradiction. Nevertheless, the characteristic of logical propositions that we are in search of is the one which was felt, and intended to be defined, by those who said that it consisted in deducibility from the law of contradiction. This characteristic, which, for the moment, we may call *tautology*, obviously does not belong to the assertion that the number of individuals in the universe is n, whatever number n may be. But for the diversity of types, it would be possible to prove logically that there are classes of n terms, where n is any finite integer; or even that there are classes of $\aleph_0$ terms. But, owing to types, such proofs, as we saw in Chapter XIII., are fallacious. We are left to empirical observation to determine whether there are as many as n individuals in the world. Among "possible" worlds, in the Leibnizian sense, there will be worlds having one, two, three, ... individuals. There does not even seem any logical necessity why there should be even one individual [43]—why, in fact, there should be any world at all. The ontological proof of the existence of God, if it were valid, would establish the logical necessity of at least one individual. But it is generally recognised as invalid, and in fact rests upon a mistaken view of existence—*i.e.* it fails to realize that existence can only be asserted of something described, not of something named, so that it is meaningless to argue from "this is the so-and-so" and "the so-and-so exists" to "this exists." If we reject the ontological argument, we seem driven to conclude that the existence of a world is an accident—*i.e.* it is not logically necessary. If that be so, no principle of logic can assert "existence" except under a hypothesis, *i.e.* none can be of the form "the propositional function so-and-so is sometimes true." Propositions of this form, when they occur in logic, will have to occur as hypotheses or consequences of hypotheses, not as complete asserted propositions. The complete asserted propositions of logic will all be such as affirm that some propositional function is *always* true. For example, it is always true that if p implies q and q implies r then p implies r, or that, if all α's are β's and x is an α then x is a β. Such propositions may occur in logic, and their truth is independent of the existence of the universe. We may lay it down that, if there were no universe, *all* general propositions would be true; for the contradictory of a general proposition (as we saw in Chapter XV.) is a proposition asserting existence, and would therefore always be false if no universe existed.

[43] The primitive propositions in *Principia Mathematica* are such as to allow the inference that at least one individual exists. But I now view this as a defect in logical purity.

Logical propositions are such as can be known *a priori*, without study of the actual world. We only know from a study of empirical facts that Socrates is a man, but we know the correctness of the syllogism in its abstract form (*i.e.* when it is stated in terms of variables) without needing any appeal to experience. This is a characteristic, not of logical propositions in themselves, but of the way in which we know them. It has, however, a bearing upon the question what their nature may be, since there are some kinds of propositions which it would be very difficult to suppose we could know without experience.

It is clear that the definition of "logic" or "mathematics" must be sought by trying to give a new definition of the old notion of "analytic" propositions. Although we can no longer be satisfied to define logical propositions as those that follow from the law of contradiction, we can and must still admit that they are a wholly different class of propositions from those that we come to know empirically. They all have the characteristic which, a moment ago, we agreed to call "tautology." This, combined with the fact that they can be expressed wholly in terms of variables and logical constants (a logical constant being something which remains constant in a proposition even when all its constituents are changed)—will give the definition of logic or pure mathematics. For the moment, I do not know how to define "tautology." [44] It would be easy to offer a definition which might seem satisfactory for a while; but I know of none that I feel to be satisfactory, in spite of feeling thoroughly familiar with the characteristic of which a definition is wanted. At this point, therefore, for the moment, we reach the frontier of knowledge on our backward journey into the logical foundations of mathematics.

[44] The importance of "tautology" for a definition of mathematics was pointed out to me by my former pupil Ludwig Wittgenstein, who was working on the problem. I do not know whether he has solved it, or even whether he is alive or dead.

We have now come to an end of our somewhat summary introduction to mathematical philosophy. It is impossible to convey adequately the ideas that are concerned in this subject so long as we abstain from the use of logical symbols. Since ordinary language has no words that naturally express exactly what we wish to express, it is necessary, so long as we adhere to ordinary language, to strain words into unusual meanings; and the reader is sure, after a time if not at first, to lapse into attaching the usual meanings to words, thus arriving at wrong notions as to what is intended to be said. Moreover, ordinary grammar and syntax is extraordinarily misleading. This is the case, *e.g.*, as regards numbers; "ten men" is grammatically the same form as "white men," so that 10 might be thought to be an adjective qualifying "men." It is the case, again, wherever propositional functions are involved, and in particular as regards existence and descriptions. Because language is misleading, as well as because it is diffuse and inexact when applied to logic (for which it was never intended), logical symbolism is absolutely necessary to any exact or thorough treatment of our subject. Those readers, therefore, who wish to acquire a mastery of the principles of mathematics, will, it is to be hoped, not shrink from the labour of mastering the symbols—a labour which is, in fact, much less than might be thought. As the above hasty survey must have made evident, there are innumerable unsolved problems in the subject, and much work needs to be done. If any student is led into a serious study of mathematical logic by this little book, it will have served the chief purpose for which it has been written.

THE END

CPSIA information can be obtained at www.ICGtesting.com
Printed in the USA
BVOW04s1640020414

349509BV00003B/569/P